Current Topics in
Tumor Cell Physiology and
Positron-Emission Tomography

Edited by
W. H. Knapp and K. Vyska

With a Foreword by O. Westphal

With 41 Figures

Springer-Verlag Berlin Heidelberg GmbH   1984

Prof. Dr. Wolfram H. Knapp
German Cancer Res. Center, Inst. f. Nuclear Medicine
6900 Heidelberg (FRG)

Prof. Dr. Karel Vyska
Inst. f. Medicine, Nuclear Res. Center
5170 Jülich (FRG)

ISBN 978-3-540-13007-9

Library of Congress Cataloging in Publication Data.
Main entry under title: Current topics in tumor cell physiology and positron-emission tomography. Bibliography:
p. Includes index. 1. Cancer cells – Adresses, essays, lectures. 2. Cell metabolism – Addresses, essays, lectures.
3. Tomography, Emission – Addresses, essays, lectures. I. Knapp, W. H. (Wolfram), 1945–   . II. Vyska, K. (Karel),
1938–   . [DNLM: 1. Tomography, Emission computed. 2. Cell transformation, Neoplastic. QZ 202 C976]
RC267.C78 1984   616.99'407   83-20199
ISBN 978-3-540-13007-9   ISBN 978-3-662-02393-8 (eBook)
DOI 10.1007/978-3-662-02393-8

Foreword

About ten years ago devices and equipment were developed for producing quantitative images *in vivo* of the distribution of substances with positron-emitting radionuclide label of very short half-life (down to only several minutes). The positron-emitting isotopes can be the most important bio-elements and are, therefore, suitable for the labeling of important physiologic substrates like sugars, amino acids etc. and their synthetic analogues. Their imaging, after distribution *in vivo,* is being performed by PET scanning systems.

It soon became apparent that positron-emission tomography could be very useful in the field of *cancer* biology, diagnostics and therapy. Transport, influx and efflux, metabolism of suitable labeled substrates around and in tumors can be followed over a period of time, thus allowing the localization of tumors and – equally important – the efficiency (or non-efficiency) of any given therapeutic regimen. For this purpose ^{13}N-labeled amino acids, especially glutamate, or ^{11}C- and ^{18}F-labeled glucose derivates were introduced and are already being recommended for the study of special tumors in experimental animals and man, such as brain tumors or osteosarcomas. Some glucose derivatives were developed which differ in their metabolic (enzymatic) fate. In this way, combined application of compounds, like [^{18}F]-2-fluoro-2-deoxy- and [^{11}C]-3-0-methyl-D-glucose are "expected to provide the most important imaging in tumor diagnostics" (M. Hatanaka).

In principle, tumor chemotherapy and other therapeutic approaches can now be followed *continuously*. Non-efficient therapy can be discontinued at an early state! "The information gained could be exploited to develop new strategies for treatment of malignant disease" (R. E. Reiman, G. Rosen et al.).

The costs of the necessary equipment – cyclotron and PET scanner – together with a highly sophisticated infrastructure, including experienced radiochemists, physicists, mathematicians and medically trained people – are, and will probably be, so high that only comparatively few patients may have a *direct* benefit, at least for the next decade to come. "However, research carried out in the course of studying such patients may provide oncologists in other settings with improved drugs, treatment regimens and the techniques necessary to evaluate their effectiveness" (R. E. Reiman, G. Rosen et al.).

This booklet provides knowledge and information about the present state of PET development in the field of cancer. Tumor biochemists and physiologists will certainly be stimulated to develop further suitable positron-emitting substrates by which certain tumor cells and normal cells can be better and better discriminated. Radiochemists will feel stimulated to improve quick and higher-yield procedures for the manufacture of such labeled materials in large enough quantities. More physicians and mathematicians will have to play their role in programming the whole process as well as in the qualitative and quantitative understanding and interpretation of the data obtained.

We are, thus, entering a field which has already proved its potentially high usefulness for the cancer patient, but it is – at the same time – open to further elaboration in almost all aspects of PET. This book is recommended to a broad scientific communitiy. Thanks to the authors for their most valuable reports!

Heidelberg, April 1983

Otto Westphal, D. Sc., M. D.

Professor, Chairman of the Management Board of the *German Cancer Research Center,* Heidelberg, and Emer. Director of the *Max-Planck-Institut für Immunbiologie,* Freiburg (FRG)

Introduction

The development of tumor-selective modalities of cancer treatment requires the exploitation of different functional properties of the tumor tissue and its normal environment. The knowledge of typical functional alterations associated with malignancy and the feasibility of their assessment act a key role in the development of therapeutic procedures as well as in the evaluation of the therapeutic efficiency.
Within the last two decades, a large scope of processes on the cellular and molecular level have been described which have been found to be characteristicly determined by malignant transformation. This progress has been attained e.g. in enzymology – reviews of great importance being given by George Weber and Sidney Weinhouse – and in the research of membrane function.
Despite of the accumulated knowledge in these fields, its applications to in-vivo diagnostic methodology has so far not been realized.
Indeed, there are severe obstacles in the in-vivo asssessment of dynamic processes which involve organic substrates: There are at least two biological compartments which have to be separately analyzed for the recording of the radioactive label. Further complication arises from the fact that 'classical' labels like C-14 or H-3 cannot be employed in vivo, since they emit beta radiation. Regarding the elements being essential constituants of organic compounds, there is only one group of isotopes applicable for in-vivo measurements, namely the positron emitters C-11, N-13, O-15, and F-18.
This fact can be taken advantage of, since the physical nature of the positron decay allows a precise quantification of radioactivity in small target volumes distant from the detector by means of positron-emission tomography. On the other hand, the employment of these nuclides is limited to a small number of centers where a cyclotron is available. The very short half life (2–120 min, 20 min for C-11) does not only require an accelerator at the site of application, it also demands special methodology of rapid syntheses for organic compounds.
In the last few years when positron cameras became commercially available, all centers having the facilities at their disposal, have taken intense efforts to make use of positron emitters for physiologic, pathophysiologic and metabolic studies in vivo. Since local alterations of physiology and metabolism are of special relevance to the brain, most work has been dedicated to this organ hitherto.

It is within the scope of this small book to illustrate by means of actual problems in cellular physiology of tumor growth, how an experimental sequence can be formed dealing with biochemistry, dynamic studies in cell culture, and finally with clinical realization by investigations in vivo using positron emitters.

Special attention is attributed to the essential role of specificly altered substrates when particular processes like membrane function are to be quantitated.

On the other hand, from the clinical point of view it is shown that even relatively non-specific informations about local metabolism or substrate availability in tumors, have already proven to be of decisive relevance in the individual therapy control.

Finally, new developments in the field of production and synthesis of short-lived radiotracers are presented.

Heidelberg, January 1984

                                                                                      The Editors

Table of Contents

List of Contributors

V. Becker, Inst. of Medicine, Nuclear Research Center, Jülich, FRG

R. S. Benua, Memorial Sloan-Kettering Cancer Center, 1275 York Ave., New York, NY 10021, USA

M. E. Bramwell, Sir William Dunn School of Pathology, South Parks Road, Oxford OX1 3RE, U. K.

J. M. F. Chamayou, Dept. of Nuclear Medicine, German Cancer Research Center, Heidelberg, FRG

L. E. Feinendegen, Inst. of Medicine, Nuclear Research Center Jülich, FRG

A. S. Gelbard, Memorial Sloan-Kettering Cancer Center, 1275 York Avenue, New York, NY 10021, USA

I. Kimmling, Inst. of Medicine, Nuclear Research Center, Jülich, FRG

W. H. Knapp, Dept. of Nuclear Medicine, German Cancer Research Center, Heidelberg, FRG

E. J. Knust, Inst. of Medical Radiation Physics and Biology, University of Essen, Essen, FRG

M. Hatanaka, Dept. of Serology and Immunology, Inst. for Virus Research, Kyoto University, Kyoto 606, Japan

F. Helus, Dept. of Radiochemistry, Radiopharmacology and Radioimmunology, German Cancer Research Center, Heidelberg, FRG

J. S. Laughlin, Memorial Sloan-Kettering Cancer Center, 1275 York Ave., New York, NY 10021, USA

H.-J. Machulla, Inst. of Medical Radiation Physics and Biology, University of Essen, Essen, FRG

W. Maier-Borst, Dept. of Radiochemistry, Radiopharmacology and Radioimmunology, German Cancer Research Center, Heidelberg, FRG

S. Matzku, Dept. of Radiochemistry, Radiopharmacology and Radioimmunology, German Cancer Research Center, Heidelberg, FRG

H. M. Mehdorn, Dept. of Neurosurgery, University of Essen, Essen, FRG

H. Ostertag, Dept. of Nuclear Medicine, German Cancer Research Center, Heidelberg, FRG

M. Profant, Dept. of Applied Mathematics, University of Duisburg, Duisburg, FRG

R. E. Reimann, Memorial Sloan-Kettering Cancer Center, 1275 York Ave., New York, NY 10021, USA

Gerald Rosen, Memorial Sloan-Kettering Cancer Center, 1275 York Ave., New York, NY 10021, USA

F. Schuier, Dept. of Neurology, University of Düsseldorf, Düsseldorf, FRG

R. v. Seggern, Inst. of Mathematics, Nuclear Research Center, Jülich, FRG

G. Spohr, Inst. of Medicine, Nuclear Research Center, Jülich, FRG

K. Vyska, Inst. of Medicine, Nuclear Research Center, Jülich, FRG

M. J. Weber, Dept. of Microbiology, University of Illinois, Urbana, Ill. 61801, USA

M. K. White, Sir William Dunn School of Pathology, South Parks Road, Oxford OX1 3RE, U. K.

S. D. J. Yeh, Memorial Sloan-Kettering Cancer Center, 1275 York Ave., New York, NY 10021, USA

Metabolic and Transport Alterations in Cells Transformed by Rous Sarcoma Virus

M. J. Weber

When cells become transformed by Rous sarcoma virus, they undergo a number of morphological, metabolic and regulatory alterations which are outlined in Table 1. These changes collectively are often referred to as "the transformed phenotype" and are characteristic, not only of cells transformed *in vitro* by Rous sarcoma virus (reviewed in Hanafusa 1977) but are often found to occur in cells transformed by many other tumor viruses and by chemical carcinogens as well as in cell cultures derived from spontaneously occuring tumors. The focus of our work has been on analysing the molecular mechanism underlying the increased rate of glucose transport and metabolism in cells transformed by Rous sarcoma virus. This biological phenomenon was first described by Hatanaka and his colleagues (1970).

1. Increased Hexose Transport is Transformation-Specific. In analyzing transformation, it is important initially to distinguish between those cellular changes which are associated with alterations in growth rate (which we term growth-state contingent) and those which change specifically with malignant transformation, regardless of the rate of cellular proliferation (which we term transformation-specific). Table 2 summarizes the results of several experiments in which the initial rate of uptake of a variety of nutrients and ions was examined in chicken embryo fibroblasts which were either density-inhibited, growing exponentially or transformed by Rous sarcoma virus. It is clear that uptake rates for all the molecules examined changed in coordination with the growth rate, being slower in cells which were density inhibited and higher in cells which were replicating. Thus, all the ions and nutrients examined display growth-state contingent changes in uptake rate. However, only the rate of glucose transport

Table 1. The Transformed Phenotype

Increased glucose transport
Increased glycolysis
Increased plasminogen activator
Rounded morphology
Disrupted cytoskeleton
Loss of fibronectin
Decreased adhesiveness
Loss of growth control
Tumorigenicity

Table 2. Relative rates of nutrient uptake by density-inhibited, normal growing and RSV-transformed chicken embryo fibroblasts[a]

	Density-Inhibited	Normal growing	Trans-formed
2-deoxyglucose	1.0	6.8	23.9
3-0-methylglucose	1.0	6.5	32.8
uridine	1.0	4.0	3.4
adenosine	1.0	2.1	n.d.[b]
thymidine	1.0	9.0	8.9
α-aminoisobutyric acid	1.0	4.5	4.5
phosphate	1.0	1.9	1.8
potassium	1.0	1.8	1.6

[a] Initial uptake rates were determined as described (Weber 1973)
[b] Not determined

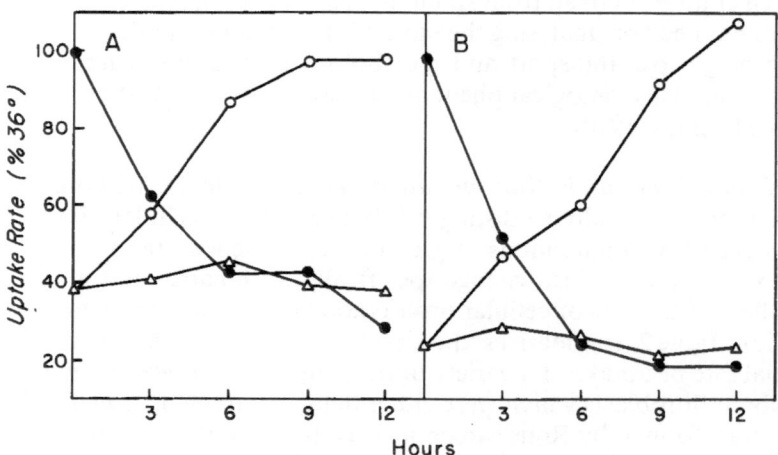

Fig. 1. Transport rates of 1mM ^{3}H-2-deoxyglucose (A) and 1mM ^{3}H-3-0-methylglucose (B) in cells infected with a temperature-conditional mutant of Rous sarcoma virus. Cells grown at 42° ($\triangle - \triangle$), shifted from 42° to 36° ($\bigcirc - \bigcirc$) or shifted from 36° to 42° ($\bullet - \bullet$). All data are expressed as a percentage of the 36° (transformed) control

changed in a transformation-specific manner: cells transformed by Rous sarcoma virus transported hexoses at a rate which was five times faster than that shown by normal cells which were proliferating at the same rate. (Note, however, that transformation-specific alterations in the *regulation* of amino acid transport have been observed (Nakamura and Weber 1979).

The notion that the the increased rate of hexose transport in Rous-transformed cells is "transformation-specific" is bolstered by the finding that the increase in transport is dependent on the continuous activity of pp60[src], the viral transforming protein. This is shown in Figure 1. Cells infected with a mutant of Rous sarcoma virus which codes for a thermolabile pp60[src] protein (Martin et al 1971;

Table 3. Kinetic parameters of hexose transport

| | Kinetic parameters of 2-deoxyglucose transport | | | | | |
| | Contact-inhibited | | Normal, growing | | Transformed | |
	Km^a	$Vmax^b$	Km	Vmax	Km	Vmax
A. 0.019–10.0mM	4.7 ±1.8	7.9 ±2.0	3.0 ±1.1	11 ±2.9	1.0 ±0.39	16 ±1.5
B. 0.019–1.0mM	0.46±0.22	1.2 ±0.24	0.52±0.23	3.0 ±0.59	0.80±0.091	15 ±1.1
C. Corrected graphically for nonsaturable uptake			0.51+0.28	1.8 +0.28	0.81+0.091	13 +1.0

| | Kinetic parameters of 3-0-methylglucose transport | | | | | |
| | Contact-inhibited | | Normal, growing | | Transformed | |
	Km^a	$Vmax^b$	Km	Vmax	Km	Vmax
A. 0.019–10.0mM	4.5 ±2.5	1.4 ±0.46	5.6 ±1.5	4.5 ±0.95	2.2 ±0.60	14 ±3.7
B. 0.019–1.25mM	0.65±0.32	0.23±0.064	0.81±0.35	0.80±0.24	0.68±0.22	5.6±1.3
C. Corrected graphically for nonsaturable uptake	2.6 ±1.6	0.56±0.22	2.2 ±1.0	2.3 ±0.53	1.5 ±0.28	10 ±1.4
D. Corrected for mannitol uptake			3.5 +0.54	0.82+0.10	2.6 +1.2	3.2+0.61

[a] Km is in millimoles
[b] Vmax is in nanomoles per min per mg of protein

Vogt 1977) display a low rate of hexose transport when held at the restrictive temperature of 42°. However, when shifted to 36°, the permissive temperature, these cells begin to increase their transport rate without detectable lag. Conversely, when the cells infected with the temperature-conditional mutant are held at 36°, the transport rate is high, but when shifted to 42°, the transport rate starts rapidly to decline.

2. Molecular Basis of the Increased Transport Rate. Several lines of evidence suggest that the increased glucose transport rate in transformed cells is due to an increased synthesis of the transporters in the transformed cells.

2.1 The increased transport is associated with an increased Vmax with no detectable, reliable change in the Km for transport (Table 3) (Weber 1973). The reason several sets of kinetic parameters are shown in Table 3 is because different procedures for measuring Km and Vmax give rise to different values for these parameters. This is a consequence of the fact that cultured cells display a considerable "non-saturable" component to their hexose uptake. The "non-saturable" uptake could be due to trapping in the extracellular space, passive diffusion, uptake by low-affinity transport systems, or a combination of these. But in any case, when care is taken to compensate for the "non-saturable" component, we find an increased Vmax for hexose transport in the Rous-transformed cells, but no significant change in Km.

Table 4. Hexose uptake and cytochalasin B binding to normal and transformed intact cells and membrane fractions[a]

Cells	Assay conditions	2-Deoxyglucose uptake	Cytochalasin B binding		
			Intact cells	Membrane-enriched fraction	Purified membranes
		pmol/mg protein/15 min	pmol/mg protein		
Trans-formed	+ D-Mannitol	8 783	42.8	41.3 ± 2.1	89.2
	+ D-Glucose	349	25.4	9.7 ± 0.6	19.4
	Glucose-specific	8 434	17.4	31.6	69.8
Normal Confluent	+ D-Mannitol	1 084	20.2	9.8 ± 0.4	18.0
	+ D-Glucose	748	18.6	6.9 ± 0.3	11.6
	Glucose-specific[b]	336	1.6	2.9	6.4

[a] D-Mannitol and D-glucose were present in the assay mixture at a final concentration of 182 mM for intact cells and membrane-enriched fractions and 200 mM for purified membranes. 2-Deoxyglucose uptake and cytochalasin B binding were performed as described in Salter and Weber (1979). Data for 2-deoxyglucose uptake and cytochalasin B binding to intact cells are the average of duplicate determinations and the individual numbers varied by less than 10%. Data for membrane enriched fractions are multiple determinations from the same samples and the numbers are average ± S. D. Data for purified membranes are single point determinations

[b] Uptake or binding in the presence of D-mannitol minus uptake or binding in the presence of D-glucose

In contrast to our results, Bramwell and his colleagues (this volume) find a decreased Km to be associated with malignancy in a panel of normal x malignant cell hybrids. Differences in biological system rather than methodology are likely to explain these divergent results. For example, if malignant cells express an embryonic glucose transporter with a low Km, no change in Km would be expected in our system, which uses chicken embryo fibroblasts.

2.2 The increased transport rate induced by the action of the Rous sarcoma virus transforming protein, pp60[src], requires new protein synthesis (Kawai and Hanafusa 1971; Kletzien and Perdue 1976).

2.3. The amount of glucose-specific cytochalasin B binding in transformed cells is increased in proportion to the increased transport rate (Salter and Weber 1979) as shown in Table 4. Cytochalasin B is a potent inhibitor of glucose transport in animal cells, and a portion of the binding of radioactive cytochalasin B to cells or to cell membranes can be inhibited by D-glucose. We term this portion of the cytochalasin B binding the "glucose specific" binding, and we believe it represents binding to the glucose transporter. It is clear that the amount

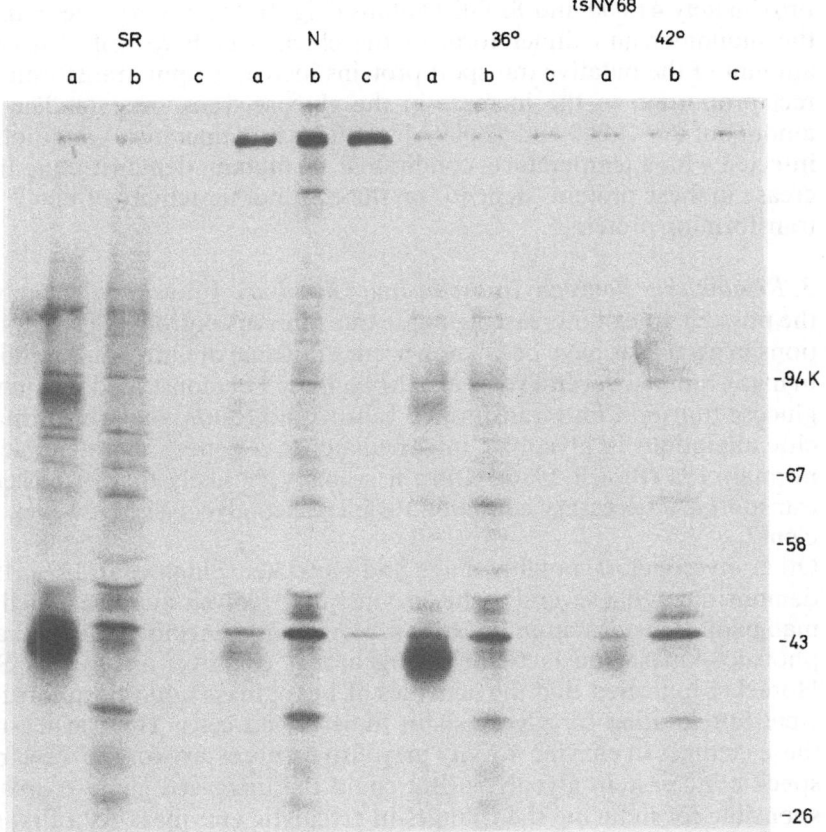

Fig. 2. Gel electropherogram of immunoprecipitates of particulate fractions from cells transformed by RSV (SR), normal cells (N), or cells infected with the temperature-conditional *src* mutant tsNY68 and held either at the permissive (36°) or non-permissive (42°) temperatures. a, immune serum; b, immune serum, blocked with excess human glucose transporter; c, pre-immune serum

of this cytochalasin B binding is increased in transformed cells. This provides strong, but still indirect evidence that increased transport of glucose in transformed cells is associated with an increase in the number of glucose transporters.

2.4. The most direct evidence that the increased transport rate is due to an increase in the number of transporters comes from work performed in collaboration with Dr. Gustav Lienhard (Dartmouth, U.S.A.). Lienhard and his collaborators have purified the glucose transporter from human erythrocytes (Baldwin et al. 1979). Using antibody raised against the human erythrocyte transporter (Baldwin and Lienhard 1980; Salter et al. 1982) we have detected cross reactive antigens in chicken embryo fibroblasts, the major ones being approximately 41,000 and 82,000 Daltons (Fig. 2). We believe, these antigens are the monomer and dimer form of the chicken embryo cell transporter. The amount of the putative transport proteins increases upon transformation in direct proportion to the increase in the glucose transport rate. The increased amount of the 41,000 and 82,000 Mr proteins is temperature conditional in cells infected with a temperature- conditional *src* mutant, demonstrating that the increase in these proteins depends on the continuous activity of pp60[src], the viral transforming protein.

3. Relationship Between Transport and Glycolysis. I now would like to discuss the possible role of increased glucose transport in controlling glycolysis. Alterations in glycolysis have been known since the time of Otto Warburg to be found in many tumors. Several years ago, Mina Bissell demonstrated that inhibition of glucose transport into transformed cells would restore the transformation-specific alterations in glycolytic intermediates to a pattern more characteristic of normal cells (Bissell 1976). Thus, it seems very likely that increased glucose transport is a necessary condition for increased glycolysis. However, is it sufficient?

Other investigators, notably Singh and Horecker (Singh et al. 1974, 1978) have demonstrated that several of the enzymes of glycolysis increase in activity upon malignant transformation by Rous sarcoma virus, including hexokinase, phosphofructokinase and lactic dehydrogenase. Cross-over analysis by Singh and Horecker indicated that the activities of hexokinase and phosphofructokinase were rate-limiting for glycolysis in transformed cells. Thus, at least some of these changes in enzyme activity may also be necessary for the transformation-specific increase in glycolysis. But could the increased glucose uptake be responsible for inducing the changes in glycolytic enzymes? As a first approach to this question, we recently have begun an investigation of the time course with which the Rous sarcoma virus *src* gene induces these changes in glycolytic enzymes. We have used cells infected with a temperature-conditional *src* mutant of Rous sarcoma virus and held at the restrictive temperature of 42°. Under these conditions, the cells are phenotypically normal and have a low rate of glucose transport and glycolysis. When the cells were shifted to 36°, they started increasing their glucose transport rate (Fig. 3) and the change in transport capacity started to occur with little apparent lag. The increase in hexokinase activity occurred no more quickly than did the change in transport rate, and may have followed that increase by as much as a few hours. The increase in lactic dehydrogenase activity occurred very slowly upon expression of the *src* gene product, and showed a lag of around 24 hours. Thus, the increased activity of this last enzyme is likely to be a secondary consequence of the other alterations in glucose uptake and metabolism rather than a primary regulatory event. These data are consistent with the notion that increased glucose transport is the primary event in causing transformed cells to alter their glycolytic metabolism.

4. Relevance to Positron Emission Tomography. The increased glucose transport rate described in this report, as well as the other markers of transformation listed in Table 1, have been examined primarily *in vitro*, particularly in cell culture.

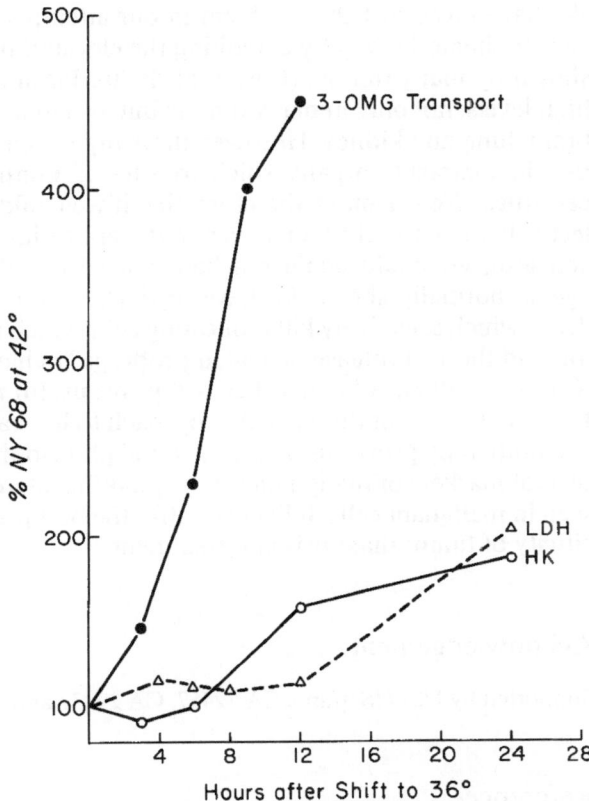

Fig. 3. Time course of increases in glucose transport rate and glycolytic enzymes during transformation by Rous sarcoma virus. Cells infected with tsNY68 (Kawai and Hanafusa, 1971) and held at 42° were shifted to the permissive temperature (36°) and at various times the rate of transport of 3-0-methylglucose and the activity of hexokinase (HK) and lactic dehydrogenase (LDH) were measured

Positron emission tomography can make an enormous contribution to our understanding of tumor physiology if it permits us to determine whether these parameters of transformation are also expressed in tumors *in situ*.

If these various transformation parameters are, in fact, expressed *in situ* by human tumors, and if they are of a sufficient magnitude relative to non-specific pharmacological factors, such as blood flow and tissue distribution, they conceivably could be useful in cancer diagnosis and therapy. However, all the various parameters of transformation are characteristic, not only of malignant cells, but also of at least some normal cells at some stage of development. That is to say, there is no such thing as a completely specific "marker" for cancer. What one sees in malignant cells are inappropriate patterns of normal cellular properties. If one wishes to find a truly specific "marker" for cancer, one may have to look for just such inappropriate combinations of cellular properties.

We have faced just this problem in our attempts to increase the specificity of cancer chemotherapy by exploiting the elevated plasminogen activator activity shown by many tumors (Carl et al. 1980). Plasminogen activator is present at high levels, not only in many tumors but also in many normal organs, including brain, lung and kidney. However, these organs have a relatively low mitotic index, in contrast to organs which are sites of limiting toxicity for many anticancer drugs. We reasoned, therefore, that if we could design a drug which was selectively toxic to cells which were *both* replicating and producing plasminogen activator, we could achieve a higher level of selectivity than anti-neoplastic agents normally show. We have, in fact, synthesized a group of anti-cancer drugs which selectively kill replicating cells which produce plasminogen activator, and these "protease-activated prodrugs" behave in cell culture as expected (Carl et al. 1980). Whether they will prove useful *in vivo* is under investigation. But whether or not this specific approach to increasing the specificity of cancer chemotherapy proves useful, the general philosophy of utilizing simultaneously several markers of malignancy, to exploit the altered patterns of differentiation seen in malignant cells, it likely to offer the best prospect for increasing the specificity of tumor diagnosis and treatment.

Acknowledgement

Supported by USPHS grants CA 12467, CA 29837 and CA 32964.

References

1. Baldwin SA, Baldwin JM, Gorga FR and Lienhard GE (1979) Purification of the cytochalasin B binding component of the human erythrocyte monosaccharide transport system. Biochem Biophys Acta 552; 183–188
2. Baldwin SA and Lienhard GE (1980) Immunological identification of the human erythrocyte monosaccharide transporter. Biochem. Biophys Res Comm 94; 1401–1408
3. Bissell MJ (1976) Transport as a rate limiting step in glucose metabolism in virus-transformed cells: Studies with cytochalasin B. J Cell Physiol 89; 701–710
4. Carl PL, Chakravarty PK, Katzenellenbogen JA and Weber MJ (1980) Protease-activated "prodrugs" for cancer chemotherapy. Proc Natl Acad Sci USA 77; 2224–2228
5. Hanafusa H (1977) Cell transformation by RNA tumor viruses. In: Fraenkel-Conrat H and Wagner RR (eds) Comprehensive Virology, Vol 10. New York, Plenum Press, pp 401–483
6. Kawai A and Hanafusa H (1971) The effects of reciprocal changes in temperature on the transformed state of cells infected with a Rous sarcoma virus mutant. Virology 46; 470–479
7. Kletzien RF and Perdue JF (1976) Regulation of sugar transport in chick embryo fibroblasts and in fibroblasts transformed by a temperature-sensitive mutant of the Rous sarcoma virus. J Cell Physiol 89; 723–728
8. Lang DR and Weber MJ (1978) Increased membrane transport of 2-deoxyglucose and 3-O-methylglucose is an early event in the transformation of chick embryo fibroblasts by Rous sarcoma virus. J Cell Physiol 94; 315–319

9. Martin GS, Venuta S, Weber M and Rubin H (1971) Temperature-dependent alterations in sugar transport in cells infected by a temperature-sensitive mutant of Rous sarcoma virus. Proc Natl Acad Sci USA 68; 2739–2741
10. Nakamura KD and Weber MJ (1979) Amino acid transport in normal and Rous sarcoma virus-transformed chicken embryo fibroblasts. J Cell Physiol 99; 15–22
11. Salter DW and Weber MJ (1979) Glucose-specific cytochalasin B binding is increased in chicken embryo fibroblasts transformed by Rous sarcoma virus. J Biol Chem 254; 3554–3561
12. Salter DW, Baldwin SA, Lienhard GE and Weber MJ (1982) Proteins antigenically related to the human erythrocyte glucose transporter in normal and Rous sarcoma virus-transformed chicken embryo fibroblasts. Proc Natl Acad Sci USA 79; 1540–1544
13. Singh WN, Singh M, August JT and Horecker BL (1974) Alterations in glucose metabolism in chick-embryo cells transformed by Rous sarcoma virus: Intracellular levels of glycolytic intermediates. Proc Natl Acad Sci USA 71; 4129–4132
14. Singh M, Singh V, August JT and Horecker BL (1978) Transport and phosphorylation of hexoses in normal and Rous sarcoma virus-transformed chick embryo fibroblasts. J Cell Physiol 97; 285–292
15. Vogt PK (1977) The genetics of RNA tumor viruses. In: Fraenkel-Conrat H and Wagner RR (eds) Comprehensive Virology, Vol 9. New York, Plenum Press, pp 341–455
16. Weber MJ (1973) Hexose transport in normal and Rous sarcoma virus-transformed cells. J Biol Chem 248; 2978–2983

Some Tumour Markers in Malignant and Non-Malignant Hybrid Cells

M. E. Bramwell and M. K. White

1. Introduction

The use of positron emission tomography is of great potential in the detection of tumors. This report reviews recent studies on particular physiological and biochemical markers that appear to have a good correlation with the malignant phenotype.

2. Cell Fusion

The study of tumour markers in cell hybrids has its origins in the work of Harris and coworkers (1). It has been shown that when a malignant mouse cell is fused, by means of a virus, with a normal mouse cell (eg. fibroblast), the resultant hybrid cell is non-tumorigenic in nude mice. That is, the normal mouse cell suppresses the malignancy of the tumour cell (malignancy in this context is defined as the ability of cells to grow progressively in immuno-suppressed or nude mice).
Subsequent cultivation of the cells *in vitro* under different conditions of growth yield revertants which are malignant. Revertants also are formed occasionally *in vivo* after a prolonged period if a large number (10^7) of supressed cells are injected.

3. Properties of Malignant Cells

Chromosome analysis of these segregants (2) revealed that loss of only a few chromosomes occurred, notably chromosome 4 (in the mouse) (3). These hybrid pairs of cells (one malignant and one non-malignant) thus provide an ideal system for examining changes that occur during malignant transformation. Many such pairs have now been constructed involving mouse-mouse crosses, mouse-human crosses (4, 5, 6) Chinese hamster – Chinese hamster (7) and human-human (8, 9). Some of the properties ascribed to malignant cells, i.e. changes in morphology, growth in low serum concentration, growth-rate characteristics, growth in soft agar and microtubule organization have been shown to be characteristic of some but not *all* malignant cells (10, 11, 12).

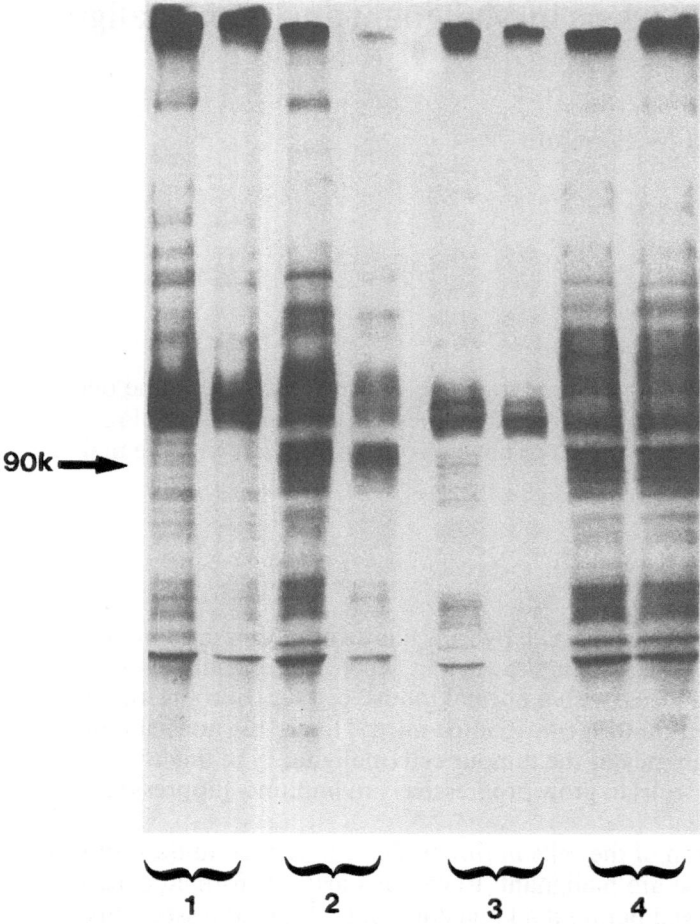

Fig. 1. Autoradiograph of a SDS-PAGE of cells labelled metabolically with [^{14}C]-glucosamine

Lane 1   C57 (mouse fibroblast)
Lane 2   PG19 (mouse melanoma)
Lane 3   PG19 × T13H (Hybrid cell) clone 8 (suppressed)
Lane 4   PG19 × T13H Clone 8T1 (malignant revertant)

4. Membrane Glycoproteins of Cell Hybrids

However, examination of membrane glycoproteins using a radio-iodinated lectin overlay technique or polyacrylamide gels led to the discovery of two glycoproteins of molecular mass approximately 100K and 90K which were apparent only in the malignant cells (13, 14, 15).

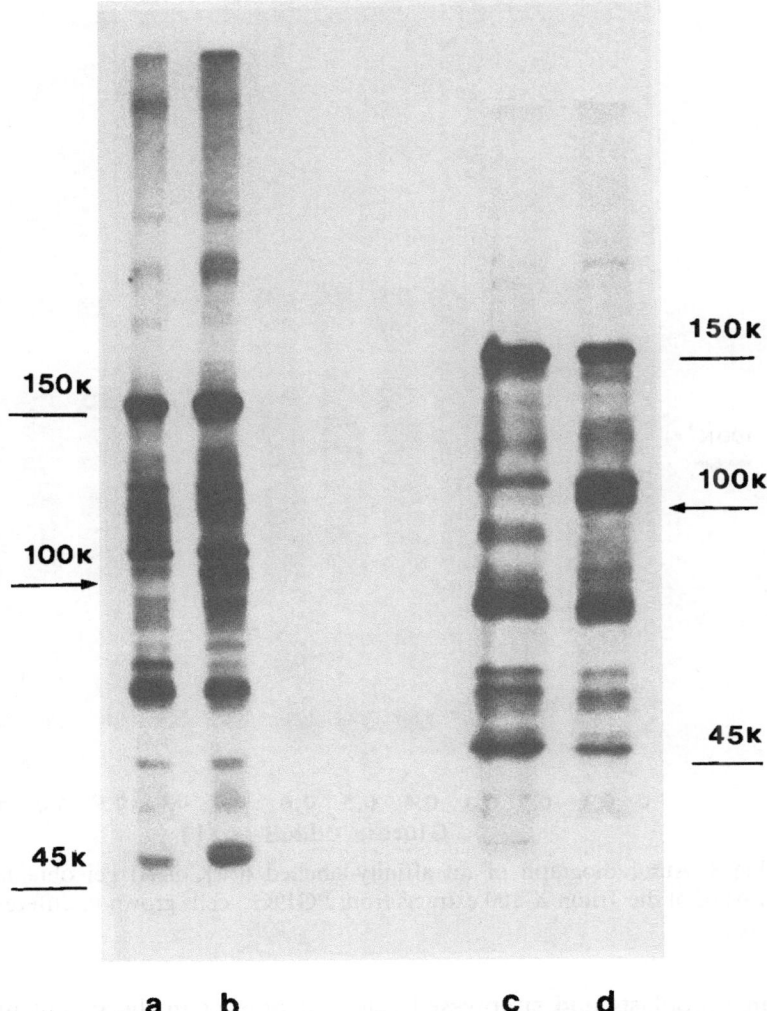

Fig. 2. Autoradiographs affinity-labelled with [^{125}I]Con A, obtained from SDS/PAGE of the Triton X-100 extracts of: a, PG19G$^-$ cells grown in low glucose levels (0.1 g/l); b, PG19G$^-$ cells grown in normal glucose levels (1 g/l); c, suppressed hybrid cell line PG 19 × human lymphocyte C119 6TG; d, revertant malignant cell line PG19 × human lymphocyte C119 6TGT1

4.1 90K Glycoprotein

Figure 1 shows the presence of the 90K band present in an autoradiograph of [^{14}C]glucosamide labelled cells. This compound preferentially labels glycoproteins which are rich in sialic acid residues. Glucosamine appears to be largely incorporated into sialic acid residues and thus a glycoprotein pattern is revealed which is very similar to that obtained by affinity labelling the gel with radioactive wheat germ agglutinin (WGA). It is clear that the 90K band is absent

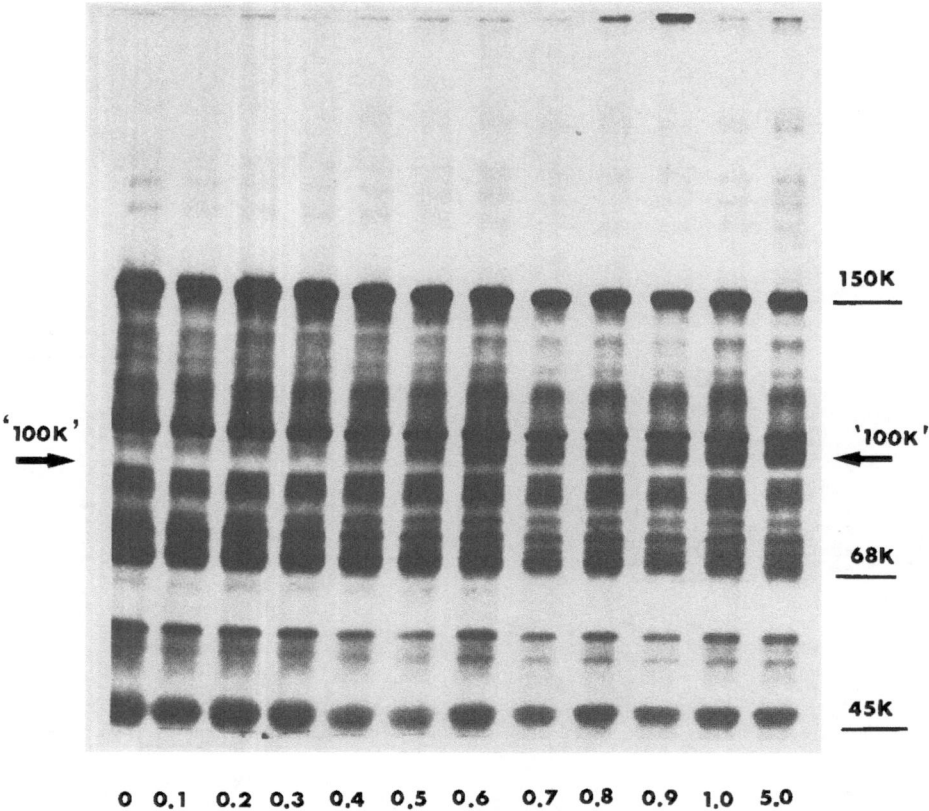

'100K'

150K

'100K'

68K

45K

0 0.1 0.2 0.3 0.4 0.5 0.6 0.7 0.8 0.9 1.0 5.0
Glucose added (g/l)

Fig. 3. Autoradiograph of an affinity-labelled ([^{125}]Con A) gel obtained from SDS/ PAGE of the Triton X-100 extracts from PG19G$^-$ cells grown in different levels of glucose

in fibroblasts and suppressed cells but present in the parent melanoma cell (PG19) and the malignant revertant.

4.2 100K Glycoprotein

If a lectin, concanavalin A is used instead to affinity label the gel, then the 100K glycoprotein can be revealed in malignant cells, Figure 2. This lectin binds strongly to glucose and mannose indicating that the 100K glycoprotein is a 'high-mannose' type. Since the use of hybrid cells has defined two glycoproteins of interest, further studies were carried out on the parent malignant melanoma cell line (PG19). A variant of this cell was obtained by selection in reduced levels of glucose and was termed PG19G$^-$ (16, 17). In this clone it was found that the 100K and 90K glycoproteins apeared to be inducible with glucose. Whether or not this effect is a true indication of glycoprotein synthesis or an effect of glycosylation is still open to question, but it was thought that these two glycoproteins might be involved in someway in glucose transport since they were the only ones significantly altered in response to changes in glucose levels (Figure 3). Subsequently it was found that the 90K glycoprotein varied in amount with glucose levels even in the wild type PG19. It was therefore concluded that these two glycoproteins were under separate control.

5. Glucose Transport

Glucose uptake changes have been implicated in malignant transformation for a very long time (18, 19, 20) and in view of the glucose sensitivity of both the 90K and 100K glycoproteins (putative markers of the malignant change), it was an obvious step to examine the kinetics of glucose metabolism of the parent cells and the hybrid cell pairs.

5.1 Kinetics in Malignant and Non-Malignant Cells

It was found somewhat surprisingly that the Vmax was independent of the malignant character of the cell, being greatly affected by growth conditions and the

Table 1. Kinetic constants of hexose uptake in malignant and normal cells

Cell type	Cell density (cells/cm^2)	Passage no.	Km (mM)	Vmax (nmoles/10 cells/h)
Malignant				
PG 19	2.5×10^5		0.608 ± 0.114	96.1 ± 7.8
	1.2×10^5		0.992 ± 0.119	108.2 ± 7.8
YACIR	suspension culture		1.101 ± 0.033	44.7 ± 0.8
A9HT	2.1×10^5		1.170 ± 0.079	131.4 ± 6.3
SEWA	suspension culture		0.683 ± 0.008	494.7 ± 2.6
TA3HaB	2.0×10^5		0.762 ± 0.078	291.2 ± 16.7
Normal				
C57Black fibroblasts	4.6×10^4	2	2.510 ± 0.232	184.0 ± 12.6
	4.8×10^4	5	2.500 ± 0.494	295.7 ± 44.7
CBAT6T6 fibroblasts	3.5×10^4	3	1.622 ± 0.182	255.2 ± 16.1
	9.9×10^4	4	1.773 ± 0.153	287.9 ± 13.2
T13HT13H fibroblasts	1.1×10^5	4	1.610 ± 0.388	162.8 ± 19.3
	7.1×10^4	4	1.730 ± 0.127	387.5 ± 15.9
Rb7BnR/Rb7BnR fibroblasts	9.4×10^4	2	2.153 ± 0.155	184.6 ± 8.6
C57Black lymphocytes	suspension culture		1.740 ± 0.085	4.5 ± 0.17

Table 2. Kinetic constants of hexose uptake in melanoma × fibroblast hybrids and in a lectin-resistant melanoma derivative

Cell type	Tumorigenicity	Cell density (cells/cm^2)	Km (mM)	Vmax (nmoles/10^6 cells/h)
PG19 × T13HT13H Clone 7	−	2.1×10^5	2.348 ± 0.339	195.0 ± 16.4
PG19 × T13HT13H Clone 8	−	5.3×10^4	3.587 ± 0.572	160.5 ± 18.4
		7.9×10^4	2.533 ± 0.466	132.5 ± 16.2
PG19 × T13HT13H Clone 8T1	+	2.1×10^5	1.480 ± 0.051	204.3 ± 3.8
PG19WGAR Clone C2	−	2.0×10^5	2.40 ± 0.230	173.5 ± 11.5

Kinetic constants of hexose uptake in fibrosarcoma × lymphocyte hybrids and in lectin-resistant fibrosarcoma derivatives

Cell type	Tumorigenicity	Cell density (cells/cm^2)	Km (mM)	Vmax (nmoles/10^6 cells/h)
A9HT × C57Black lymphocyte C13	−	2.6×10^5	2.240 ± 0.160	247.6 ± 13.2
A9HT × C57Black lymphocyte C14	−	2.5×10^5	2.080 ± 0.120	451.3 ± 20.8
A9HT × C57Black lymphocyte C12T1	+	2.0×10^5	1.123 ± 0.046	119.4 ± 2.8
A9HT WC	−	1.8×10^5	1.685 ± 0.151	133.9 ± 7.6
A9HT WD	+	1.8×10^5	0.731 ± 0.087	120.4 ± 7.9

degree of confluency of the cell population. However, it became clear that the Km for uptake was consistently reduced in malignant cells (21) (Tables 1 and 2). This indicates an increased affinity for glucose by the transporter. The uptake studies were performed largely using 2-deoxyglucose, a glucose analogue which is not metabolized beyond the first phosphorylation step. Studies were also carried out using 3-0-methyl glucose which is not phosphorylated and considerations of the effect of hexokinase on transport can, therefore, be excluded (Table 3). It is also apparent that 3-0-methyl glucose has a lower affinity than 2-deoxyglucose for the transporter and it emphasizes more clearly the difference between suppressed and malignant cells.

5.2 Murine Melanoma Cell Line

Because the mouse melanoma clone PG19G$^-$ showed inducibility for the 90K and 100K, it was of interest to measure the Vmax and Km of this cell line in conditions of normal and low glucose. Table 4 shows a comparison of the effect

Table 3. Kinetic parameters of 3-0-methyl-D-glycose uptake in malignant and non-malignant cells

Cell type	Vmax	Km
CBAT$_6$T$_6$ fibroblast	325.1 ± 26.8	4.4 ± 0.5
	420.3 ± 90.8	4.8 ± 1.2
PG19 (Melanoma)	106.6 ± 5.1	2.1 ± 0.1
	116.6 ± 18.2	2.2 ± 0.4
1Acn2 (Suppressed)	331.2 ± 40.2	8.4 ± 1.2
1Acn1TG (Malignant)	107.6 ± 7.3	2.7 ± 0.2

Table 4. Effect of glucose starvation on the kinetic parameters of deoxyglucose uptake by tumorigenic cells

Cell type	+ glucose				− glucose			
	Cell density (cells/cm^2)	n	Km (mM)	Vmax (nmoles/106 cells/h)	Cell density (cells/cm^2)	n	Km (mM)	Vmax (nmoles/106 cells/h)
PG19	1.2 × 10^5	6	0.99 ± 0.12	108.2 ± 7.8	2.0 × 10^5	6	0.85 ± 0.10	239.6 ± 15.2
PG19-G$^-$	1.5 × 10^5	6	0.84 ± 0.05	50.7 ± 1.8	2.5 × 10^5	6	1.82 ± 0.14	383.2 ± 17.9

n = number of concentrations assayed

of low levels of glucose on the clone PG19G$^-$ and the wild type parent cell PG19. It is clear that with PG19G$^-$ in low levels of glucose both the Km and Vmax are raised, but that in PG19 itself only the Vmax is affected. This is good circumstantial evidence that the transporter itself or a regulatory protein associated with it, is altered in the malignant state. Increases in Vmax are generally understood to represent an increase in the number of transporters present on the cell membrane (22) so that with both PG19 and PG19G$^-$ there is an increase in the number of transporters on glucose starvation, but in PG19G$^-$ only these have an altered Km which approaches a 'normal' level (i. e. non-malignant) and cease expressing the 90K and 100K glycoproteins.

One explanation for the alteration of Km is that the degree of glycosylation of the protein affects its affinity for glucose, and this hypothesis is supported by the effect of tunicamycin, an antibiotic which inhibits the glycosylation of asparagine in particular (23). Incubation of cells in tunicamycin has the effect of shifting the Km to a higher value, i. e. towards the value for normal cells (suppressed hybrids of fibriblasts) (24).

6. Monoclonal Antibodies to Membrane Glycoproteins

6.1 100K

Using a purified preparation of the 100K glycoprotein, a series of monoclonal antibodies was raised and one in particular was used to study its binding to cells *in vitro* and *in vivo* (25, 26).

It was concluded that the number of binding sites increased with glucose star-vation and was also cell cycle dependent, further evidence that the 100K was in-volved in glucose transport and that the increased binding of concanavalin A to the 100K reflected an abnormality in glycosylation which in turn affected the Km of the transporter.

6.2 90K

Using a preparation of glycoproteins which bound to a WGA-sepharose affini-ty column (i. e. those rich in sialic acid terminal groups), a series of monoclonal antibodies was generated non of which were found to be directed against the 90K glycoprotein. However, one monoclonal showed very interesting proper-ties and proved capable of discriminating between suppressed and malignant hybrid cells and between a range of tumor cells and normal cells (27). It has also proved useful in distinguishing between normal and malignant tissue in biopsy specimens (28, 29).

The antigen determinant recognized appears to be carried by a pair of high mo-lecular weight glycoproteins, rich in sialic acid of apparent molecular mass 400K and 360K.

7. Conclusion

The powerful technique of cell fusion has been used to screen for differences between normal and malignant cells; the following points have so far emerged that seem to correlate with malignancy.
1. An increase in the amount of a glycoprotein of molecular mass 100K and rich in mannose or glucose.
2. An increase in the amount of a glycoprotein of molecular mass 90K and rich in sialic acid.
3. A decrease in the Km for glucose transport.
4. The presence of a mucin-like antigen on human tumour cells of apparent mo-lecular mass 400K and 360K.

The use of glucose or glucose analogues in the detection of turnover by positron emission tomography is, therefore, justified on physiological grounds by the data discussed above. The use of a radioactively labelled tumor specific anti-body is also a most promising approach using the same technique of detection.

References

1. Harris H, Miller OJ, Klein G, Worst P and Tachibana T (1969) Suppression of malig-nancy by cell fusion. Nature (Lond) 223; 363–368
2. Jonasson J, Povey S and Harris H (1977) The analysis of malignancy by cell fusion. VII. Cytogenetic analysis of hybrids between malignant and diploid cells and tu-mours derived from them. J Cell Sci 24; 217–254

3. Evans EP, Burtenshaw MD, Brown BB, Hennion R and Harris H (1982) The analysis of malignancy by cell fusion. IX. Re-examination and clarification of the cytogenetic problem. J Cell Sci 56; 113–130
4. Weiner F, Klein G and Harris H (1971) The analysis of malignancy by cell fusion. III. Hybrids between diploid fibroblasts and other tumour cells. J Cell Sci 8; 681–692
5. Weiner F, Klein G and Harris H (1974) The analysis of malignancy by cell fusion. V. Further evidence of the ability of normal diploid cells to suppress malignancy. J Cell Sci 15; 177–183
6. Jonasson J and Harris H (1977) The analysis of malignancy by cell fusion. VIII. Evidence for the intervention of an extra-chromosomal element. J Cell Sci 24; 255–263
7. Sager R and Kovac PE (1978) Genetic analysis of tumorigenesis. 1. Expression of tumor-forming ability in hamster hybrid cell lines. Somatic Cell Genetics 4; 375–392
8. Stanbridge EJ and Wilkinson J (1978) Analysis of malignancy in human cells: malignant and transformed phenotypes are under separate genetic control. Proc Nat Acad Sci 75; 1466–1469
9. Klinger HP (1980) Suppression of tumorigenicity in somatic cell hybrids. L Suppression and re-expression of tumorigenicity in diploid human x D98-AH-2 hybrids which also demonstrate independent segregation of the tumorigenic from other cell phenotypes. Cytogenet Cell Genet 27; 254–266.
10. Straus DS, Jonasson J and Harris H (1977) Growth in vitro of tumour cell x fibroblast hybrids in which malignancy is suppressed. J Cell Sci 25; 73–86
11. Der CJ and Stanbridge EJ (1978) Lack of correlation between the decreased expression of cell surface LETS protein and tumorigenicity in human cell hybrids. Cell 15; 1241–1251
12. Watt FM, Harris H, Weber K and Osborn M (1978) The distribution of actin cables and microtubules in hybrids between malignant and non-malignant cells and in tumours derived from them. J Cell Sci 32; 419–432
13. Bramwell ME and Harris H (1978 a) An abnormal membrane glycoprotein associated with malignancy in a wide range of different tumours. Proc R Soc B 201; 87–106
14. Bramwell ME and Harris H (1978 b) Some further information about the abnormal membrane glycoprotein associated with malignancy. Proc R Soc B 203; 93–99
15. Atkinson MAL and Bramwell ME (1981) Studies on the surface properties of hybrid cells. III. A membrane glycoprotein found on the surface of a wide range of tumour cells. J Cell Sci 48; 147–170
16. Bramwell ME (1980) A glycoprotein abnormality associated with malignancy. Biochem Soc Trans 8; 697–698
17. Bramwell ME and Atkinson MAL (1982) The apparent inducibility of tumour marker glycoproteins in a melanoma cell line selected for growth in low levels of glucose. J Cell Sci 54; 241–254
18. Warburg OH (1926) Über den Stoffwechsel der Tumoren. Berlin, Springer
19. Warburg OH (1956) On the origin of cancer cells. Science 123; 309–314
20. Huebner RJ, Hatanaka M and Gilden RV (1969) Alterations in characteristics of sugar uptake by mouse cells transformed by murine sarcoma viruses. J Natl Cancer Inst 43; 1091–1096
21. White MK, Bramwell ME and Harris H (1981) Hexose transport in hybrids between malignant and normal cells. Nature Lond 294; 232–235
22. Salter DW and Weber M (1979) Glucose specific cytochalasin B binding is increased in chicken embryo fibroblasts transformed by Rous Sarcoma virus. J Biol Chem 254; 3554–3561
23. Struck DK and Lennarz WJ (1977) Evidence for the participation of saccharide-

lipids in the synthesis of the oligosaccharide chain of ovalbumin. J Biol Chem 252; 1007–1013

24. White MK (1982) D. Phil. Thesis, University of Oxford

25. Gingrich RD, Wouters M, Bramwell ME and Harris H (1981 a) Immunological definition in normal and malignant cells of a membrane protein involved in glucose transport. I. Preparation and properties of the antibody. J Cell Sci 52; 99–120

26. Gingrich RD, Wouters M, Bramwell ME and Harris H (1981 b) Immunological definition in normal and malignant cells of a membrane protein involved in glucose transport. II. Function of the antigen. J Cell Sci 52; 121–135

27. Ashall F, Bramwell ME and Harris H (1982) A new marker for human cancer cells. I. The Ca antigen and the Cal antibody. Lancet II; 1–6

28. McGee JO'D, Woods JC, Ashall F, Bramwell ME and Harris H (1982) A new marker for human cancer cells. 2. Immunohistochemical detection of the Ca antigen in human tissues with the Cal antibody. Lancet II; 7–10

29. Woods JC, Spriggs AI, Harris H and McGee JO'D (1982) A new marker for human cancer cells. 3. Immunocytochemical detection of malignant cells in serous fluids with the Cal antibody. Lancet II; 512–514

D-Glucose Transport in Oncogenic Transformation

M. Hatanaka

Introduction

When mouse embryonic fibroblasts were infected with a mouse sarcoma virus, the enhanced glucose uptake was observed by the transformed cells (Hatanaka et al. 1969).

Since then, the increased sugar uptake by transformed cells were reported from many groups (reviewed, Hatanaka 1974). The uptakes of D-mannose, D-glucose, and 2-deoy-D-glucose, beside D-glucose, also increased during transformation of cells. However, D-fructose, D-ribose, or sucrose did not show the increased uptakes (Hatanaka et al. 1970).

Table 1 shows the typical example of increased sugar uptakes by the chick embryo fibroblasts infected by Rous sarcoma virus. In general, D-glucose is taken

Table 1. Enhanced sugar uptake in chicken cells infected with RSV[a]

Substrate	Sugar uptake ratio to control	
	20 Hours p.i.	45 Hours p.i.
D-Glucose	1.3	14.7
D-Mannose	1.6	12.9
2-Deoxyglucose	1.3	8.0
D-Glucosamine	1.7	13.7
D-Galactose	1.3	3.0
3-O-Methylglucose	1.0	1.0
D-Fructose	0.9	1.4
Sucrose	0.9	0.7
Inulin	1.0	1.0
L-Leucine	1.1	1.0
Uridine	1.0	1.5
Thymidine	0.8	1.2

[a] Thorougly washed cultures were incubated with 2 ml of isotope ($\sim 10^{-6}$ M) for 10 min at 39° in a CO_2 incubator. After incubation, cells were washed thoroughly with water, scraped into distilled water, and homogenized; aliquots were taken for liquid scintillation counting and protein determinations. The cells at 20 hours p.i. had a few percent of transformants, and the cells at 45 hours p.i. were completely transformed. The value is shown as the ratio of radioactivity in RSV-infected cells compared to uninfected cells

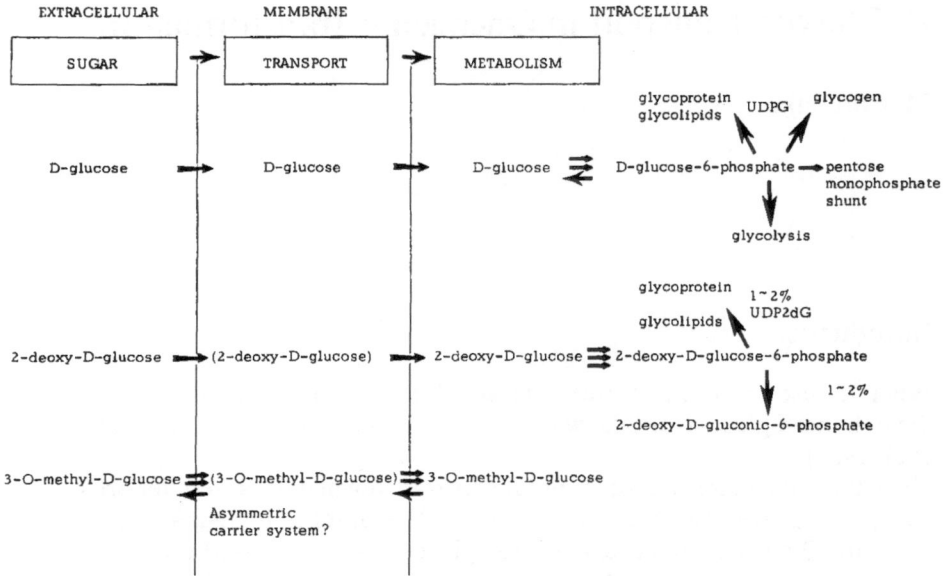

Fig. 1. Transport of D-glucose and analogs across the plasma membrane

up by the cells through the carrier transport system located in the plasma membrane. Fibroblasts, adipose cells and red blood cells reveal the facilitated diffusion for the sugar transport, while brush border membrane in kidney cortex and intestinal microvillus membrane contain the active transport system.

Facilitated diffusion of D-glucose features passive transport, saturable process at high concentration of the substrate, stereospecificity, competition with glucose analogs, and specific inhibition by cytochalasin B. D-glucose is metabolized in the cells in many pathways as shown in Fig. 1.

Labeled 2-deoxy-D-glucose and 3-O-methyl-D-glucose are used as substrates to measure the D-glucose transport activity of plasma membranes, separating from the metabolic activities and sugar pools inside the cells.

Although 2-deoxy-D-glucose is phosphorylated at 6-position by hexokinase in cells, the product, 2-deoxy-D-glucose-6-phosphate is not further metabolized in cells, except hepatic cells which contain D-glucose-6-phosphate dephosphatase activity (Hatanaka 1975).

Therefore 2-deoxy-D-glucose is taken up in linear fashion for an experimentally measurable period, 10 to 20 minutes. On the other hand, 3-O-methyl-D-glucose is not metabolized at all in cells and reaches saturation within the order of seconds which causes difficult measuring of initial rates of the transport by cells, particularly by tumors (Graff et al. 1973).

The Change of D-Glucose Transport in the Transforming Process

The change of the transport system was followed during the transforming process of chick embryo fibroblasts by Rous sarcoma virus in which system 100%

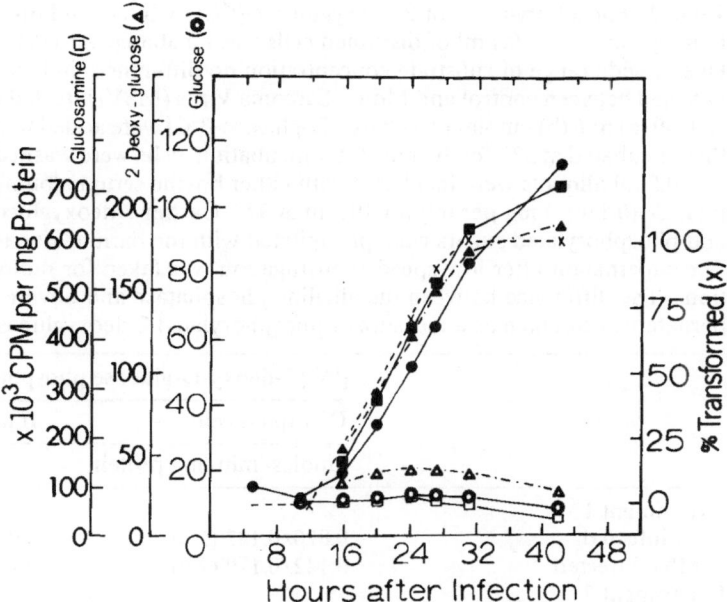

Fig. 2. Rate of uptake of some sugars into RSV-infected cells. At the indicated time after infection with RSV, cultures were pre-washed with Hanks salt solution minus glucose and incubated with 2 ml of D- 1-[^{14}C]glucose (0.2 uCi/ml, 220 Ci/mole), 2-deoxy-D-1-[^{14}C]glucose (0.2 uCi/ml, 2.3 Ci/mole) or D- 6-[^{3}H]glucosamine (2.0 uCi/ml, 260 Ci/mole), for 10 min at 39 °C. Then, samples were washed thoroughly with water, scraped into distilled water, homogenized, and aliquots taken for liquid scintillation counting and protein determination. The fraction of morphologically altered cells was estimated under the microscope. □, ○ and △, uninfected; ■, ● and ▲, RSV-infected. Reproduced with permission, from Hatanaka, M. and Hanafusa, H. (1970) Virology 41, 647–652

of the cells transform within 30 hours after infection (Hatanaka and Hanafusa 1970). As shown in Fig. 2, clear correlation was first observed between the increased D-glucose transport and transformation. Also, it became evident by measuring the transport and the phosphorylation from intact and disrupted cells, that the rate limiting reaction of D-glucose uptake into the cells is the transport system and not the phosphorylating step (Hatanaka et al. 1970) (Table 2).

It is important to analyze the kinetics of glucose transport to understand the mechanism of the enhanced sugar uptake by the transformed cells.

In the studies carried out with intact cells, the first evidence of an increased Vmax of 2-deoxy-D-glucose transport by tumor cells was obtained by using a mouse sarcoma virus. Vmax from the tumor virus infected cells was 83.5 (nmoles/mg cell protein per min), compared to 10.2 from control cells (Hatanaka et al. 1970). Since then, there has been no controversy about the increased Vmax of D-glucose transport on any of the transformed cells tested.

Table 2. Phosphorylation of 2-deoxyglucose-[^{14}C] by intact and disrupted cells. For the homogenate assay, 0.1 ml of disrupted cells was incubated with 0.4 ml of assay mixture. Over a wide range of substrate concentration no differences in hexokinase activity was detected between control and Mouse Sarcoma Virus (MSV) infected cells. For the intact, cell, 40 (a) or 6 (b) nmoles of 2-deoxy-D-glucose-[^{14}C] were added to cultures which were then incubated at 37° for 10 min. After incubation, cells were washed and homogenized and 0.5-ml aliquots were incubated with either bovine serum albumin or alkaline phosphatase (0.1 ml, 1 mg per ml) for 10 min at 37°. Carrier 2-deoxyglucose was then added and phosphorylated sugars were precipitated with the barium-zinc reagent of Somogyi. The supernatant after low speed centrifugation was taken for radioactivity determinations. The difference between the alkaline phosphatase and bovine serum albumin supernatants was taken as a measure of phosphorylated 2-deoxyglucose

	[^{14}C] 2-deoxy-D-glucose phosphorylated	
	Disrupted cells	Intact cells
	nmoles/min/mg protein	
Experiment 1		
Uninfected	0.820/0.142 (5.8)	0.024[a]/0.397 (0.06)
MSV-infected	0.142/0.179 (7.9)	0.299 /0.555 (0.54)
Experiment 2		
Uninfected	0.150/0.062 (2.4)	0.003[b]/0.410 (0.0008)
MSV-infected	0.300/0.130 (2.3)	0.036 /0.532 (0.068)

However, two basically conflicting sets of results on the Km have been found (reviewed, Hatanaka 1974).

Recently, the decrease of the apparent Km value of 2-deoxy-D-glucose transport by tumor cells has been reported by using cell fusion techniques between normal and tumor cells (White et al. 1981). It is also reported that the quantity of sugar transport carrier increased after oncogenic transformation (Salter et al. 1982; Invi et al. 1980).

When the D-glucose was omitted from the media, cells increase transport activity (Martineau et al. 1972; Hatanaka 1973). This regulation occurred at the transcriptional level, while the increased transport of sugar was shown at the post-transcriptional level (Kletzien and Perdue 1976), demonstrating that increased sugar transport by glucose starvation and transformation arises by different mechanisms.

3-O-Methyl-D-Glucose

The 3-O-methyl-D-glucose uptake system of the transformed cells demonstrated a temperature dependence (Fig. 3).

It was also found that the saturation levels of 3-O-methyl-D-glucose in the cells were directly proportional to the external concentration (Fig. 4).

Further experiments demonstrated that the rate of transport out of the cells (efflux rate) was very similar to the rate of entry into the cells (influx rate) (Graff et al. 1973).

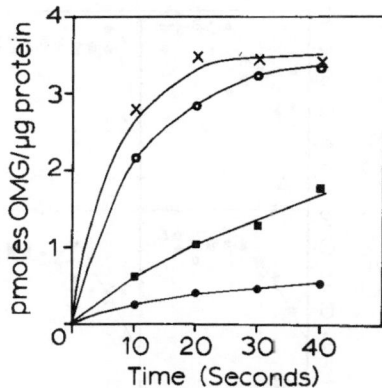

Fig. 3. The effect of temperature on the rates of 3-O-methylglucose uptake by sarcoma-virus-transformed cells. The cells were incubated with 5×10^{-4} M radioactive 3-O-methylglucose. (●), 0 °C; (■), 10 °C (○), 23 °C and (×), 37 °C. Reproduced with permission, from Graff, J. C., Hanson, D. and Hatanaka, M. (1973) Int. J. Cancer 12, 602–612

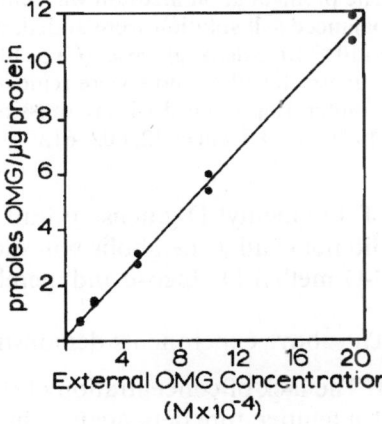

Fig. 4. Concentration dependence of the saturation levels of 3-O-methylglucose uptake by sarcoma-virus-transformed cells. The cells were incubated with radioactive 3-O-methylglucose for 10 min at 37 °C to ensure internal saturation. A linear relationship was obtained (correlation coefficient = 0.998) between the amount of sugar taken into the cell and the external sugar concentration, suggesting that the uptake process is based on diffusion. Reproduced with permission, from Graff, J. C., Hanson, D. and Hatanaka, M. (1973) Int. J. Cancer 12, 602–612

The efflux rate of 3-O-methyl-D-glucose is shown in Fig. 5. It demonstrates an accelerated efflux rate from preloaded cells exposed to medium containing an analogously transported substrate. It also shows that the two sugars used (3-O-methyl-D-glucose and 2-deoxy-D-glucose) are transported at least in part by the same carrier in the transport system.

No unique differences were found between the sugar metabolism of the control and transformed cell lines. There was neither phosphorylation nor metabolism

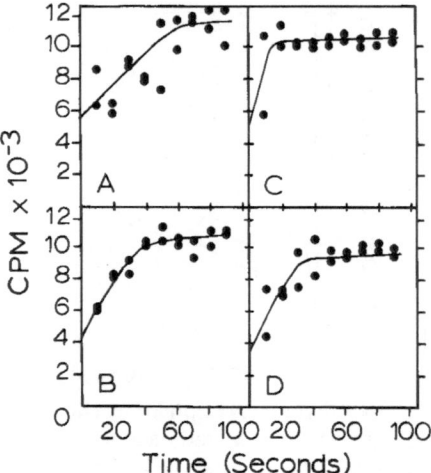

Fig. 5. The efflux rate of 3-O-methylglucose from sarcoma virus-transformed cells. The cells were preloaded with $5 \cdot 10^{-4}$ M radioactive 3-O-methylglucose for 30 min at 23 °C. The preincubation medium was removed by vacuum and immediately 3.0 ml of Hank's balanced salt solution were added, which contained: (A) no glucose or other sugar, (B) $5 \cdot 10^{-5}$ M 2-deoxyglucose, (C) $5 \cdot 10^{-3}$ M 2-deoxyglucose and (D) $5 \cdot 10^{-6}$ M 2-deoxyglucose. 100 µl aliquots were removed at intervals and counted in a liquid scintillation counter. Reproduced with permission, from Graff, J. C., Hanson, D. and Hatanaka, M. (1973) Int. J. Cancer 12, 602–612

of 3-O-methyl-D-glucose (Graff et al. 1973). Thus, oncogenic transformation did not change metabolism in such a manner as to allow the cells to metabolize 3-O-methyl-D-glucose and thereby influence the uptake rate.

The above experiments demonstrate:

1. The lack of concentration of the sustrate by the cell;
2. a temperature dependency that was indicative of a protein mediated system;
3. an efflux rate that was equivalent to the uptake rate (a property of a reversible carrier) and;
4. an accelerated efflux rate from preloaded cells exposed to medium containing an analogously transported substrate.

Thus, we conclude that 3-O-methyl-D-glucose is transported by the cells using a carrier mediated facilitated diffusion system. Together with the results of the 2-deoxy-D-glucose and 3-O-methyl-D-glucose, it is clearly demonstrated that oncogenic transformation causes a modification in the transport of hexoses. The enhanced sugar transports by transformed cells may be exploitable for early detection of some malignant tumors. We reported that 2-fluoro-2-deoxy-D-glucose, 2-deoxy-D-glucose and D-mannose inhibited oncogenic transformation by Harvey strain of murine sarcoma virus (Hatanaka 1970).

Other halogenated sugars such as 2-dichloro-2-deoxy-D-glucose or 3-fluoro-3-deoxy-D-glucose were not effective suggesting that these compounds were not sufficiently incorporated into the cells. Thus, it is concluded that [^{18}F] 2-flu-

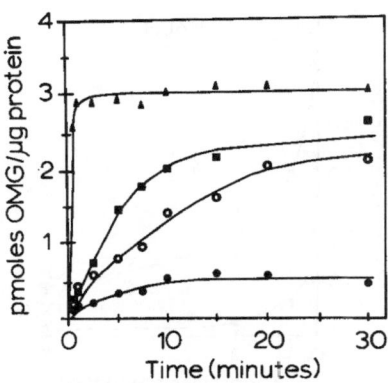

Fig. 6. The rates of 3-O-methylglucose (OMG) uptake in the presence or absence of cytochalasin B. The cells were incubated with $5 \cdot 10^{-4}$ M radioactive 3-O-methylglucose. Transformed cells without cytochalasin B (▲ — ▲), transformed cells with 10 ug/ml cytochalasin B (○ — ○), non-transformed cells with cytochalasin B (● — ●) and non-transformed cells without cytochalasin B (■ — ■). The data points represent the average of two experiments. Reproduced with permission, from Graff, J.C., Hanson, D. and Hatanaka, M. (1973) Int. J. Cancer 12, 602–612

oro-2-deoxy-D-glucose is one of the best compounds for detection of tumor *in vivo* by positron emission tomography (PET) (Hatanaka 1982). Another promising sugar compound is [^{11}C] 3-O-methyl-D-glucose. This compound was most effectively used for assessment of local perfusion and D-glucose transport rate in brain and myocardium by PET (Vyska et al.1982).

Together with our basic findings of increased transport of 3-O-methyl-D-glucose by tumors and Vyska's application of the sugar to PET, the use of [^{11}C]- 3-O-methyl-D-glucose is expected to provide the most important imaging in tumor diagnostics.

Cytochalasin B Inhibition

Sugar uptake in cells in culture is inhibited by the mold metabolite cytochalasin B (Graff et al. 1973). The effect of cytochalasin B on the uptake of 3-O-methyl-D-glucose in the parental and transformed cell can be seen in Fig.6.

Uptake by the untreated transformed cells reached saturation in less than one minute while the untreated parental cells took 15 times longer to reach saturation. Cytochasin B inhibited the initial uptake of 3-O-methyl-D-glucose into transformed cells which occurred at a rate almost equal to that of the untreated non-transformed cells, despite the presence of the inhibitor. In the presence of cytochalasin B, the parental cells showed a greatly reduced level of 3-O-methyl-D-glucose uptake that remained in effect over a long period of time, showing no sign of infiltration. Cytochalasin B was found to inhibit uptake of 2-deoxy-D-glucose in both cell types (Fig. 7). In untreated cells the uptake of 2-deoxy-D-glucose, at 5×10^{-4}M, was linear for both cell types for at least 10 min, at 37 °C (Hatanaka et al. 1970).

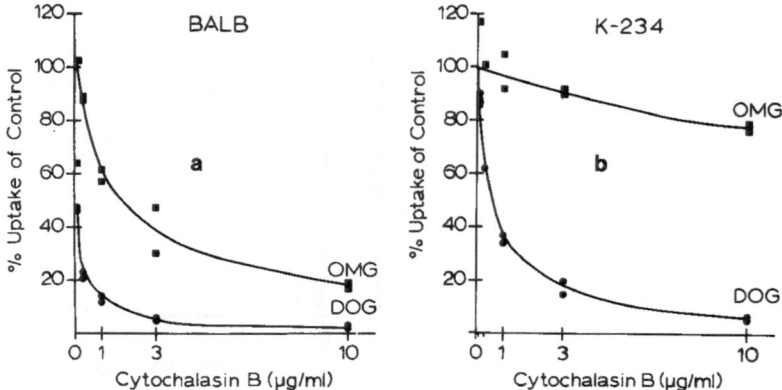

Fig. 7. Hexose uptake in the presence of various concentrations of cytochalasin B. The cells were incubated with 5×10^{-4} M radioactive 2-deoxyglucose (●) or radioactive 3-O-methylglucose (■) for 10 min at 37 °C in the presence of the appropriate concentrations of cytochalasin B, or in the case of the controls, 1% dimethylsulfoxide. (A) Non-transformed cells. (B) Transformed cells. Reproduced with permission, from Graff, J. C., Hanson, D. and Hatanaka, M. (1973) Int. J. Cancer 12, 602–612

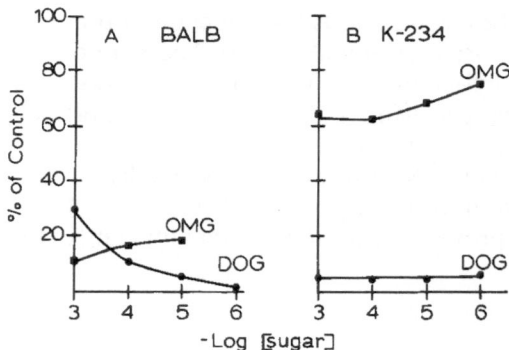

Fig. 8. Hexose uptake in the presence of cytochalasin B at various concentrations of the sugars. The cells were incubated with radioactive 2-deoxyglucose (●) or radioactive 3-O-methylglucose (■) for 10 min at 37 °C in the presence of 10 ug/ml of cytochalasin B, or in the case of the controls, 1% dimethylsulfoxide. (A) Non-transformed cells. (B) Transformed cells. Reproduced with permission, from Graff, J. C., Hanson, D. and Hatanaka, M. (1973) Int. J. Cancer 12, 602–612

The cytochalasin B inhibition of the uptake of 3-O-methyl-D-glucose was also measured at 10 min, which, on the basis of (Fig. 6), is the time in which the transformed cells was saturated with 3-O-methyl-D-glucose and the non-transformed cell was approaching saturation. Nevertheless, the experiment in Fig. 8 demonstrates that the 3-O-methyl-D-glucose uptake in transformed cells was much less inhibited than the uptake of the same sugar by the normal cell with similar amounts of cytochalasin B.

Cytochalasin B inhibition of uptake over a large range of concentrations of the two sugars was also investigated (Fig. 8). Marked differences in cytochalasin B inhibition as described above were maintained between the two cell lines over a 10,000 fold concentration range.

Therefore, 3-O-methyl-D-glucose "infiltrated" into the transformed cells, but not into the non-transformed control cells, in the presence of cytochalasin B. These observations are in contrast to those found with 2-deoxy-D-glucose uptake which was almost completely inhibited in both cells at equivalent concentrations of cytochalasin B. The term "infiltration" was used to describe incomplete inhibition of the overall uptake process, thus avoiding any inference concerning the physical nature of the membrane processes involved; and, more importantly, to describe the uptake phenomenon over a prolonged period independent of initial rate measurements. Although the infiltration observed in transformed cells, and not in control cells, cannot be explained by known mechanisms, it is possible that selective damage or specific alteration could take place in membranes of transformed cells.

Nonsaturable Process of D-Glucose Transport

The transport assays mentioned above were performed with low concentration of the substrate (10^{-3} to 10^{-7}M) and high temperature (37 °C) compared to room temperature. The assay conditions minimized the corrections for the nonsaturable process, so that most rate functions exhibit the simple hyperbolic substrate concentration curve found for the Michaelis-Menten type of kinetics known as the saturable process. Thus, the previous findings have dealt almost exclusively with the change in the saturable process with definite Km and Vmax. However, it must be noted that animal blood contains 100 mg% or 5×10^{-3}M of D-glucose, a concentration which appears to be a transitional point from saturable to nonsaturable process with physiological efficiency. The nature of the nonsaturable process of D-glucose is uncertain. It is generally assumed to represent "simple diffusion" through the lipophilic area of the plasma membrane (Plagemann and Richey 1974).

This view may be partly supported when the substrates have high lipophilicity (Zylka and Plagemann 1975). However, it is very difficult to explain the transport of less lipophilic substrates such as sugars at high concentration by "simple diffusion" for the following reasons. First, the extremely restricted uptake of highly concentrated L-glucose by the cells (Graff et al. 1973) is in sharp contrast to the high transport at high concentration (3mM) of 3-deoxy-D-glucose (Kletzien and Perdue 1974). Secondly, although simple diffusion by Lieb and Stein's model (1971) predicts that larger molecular weight compounds permeate more rapidly than lower molecular weight compounds, no increase in the rate of simple diffusion of sucrose or disaccharide was observed compared with monosaccharides (Graff et al. 1973). Therefore the nonsaturable D-glucose transport at high concentration may not be the "simple diffusion" or a simple physical process that one might imagine.

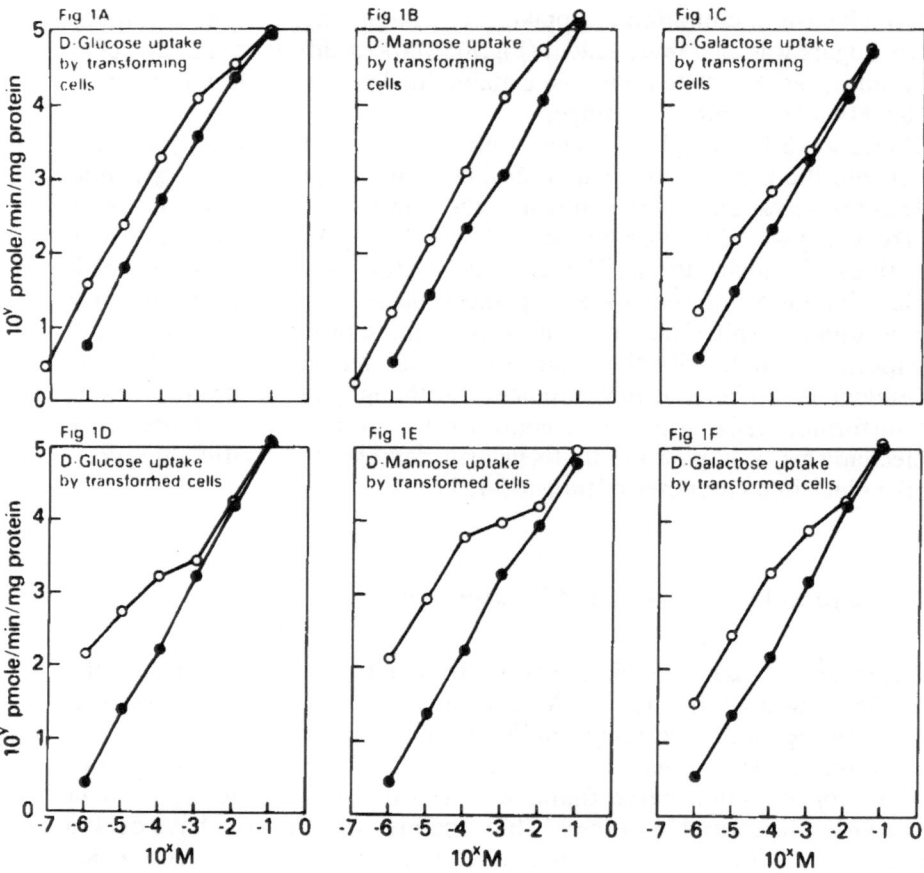

Fig. 9. Uptake of D-glucose, D-mannose and D-galactose by the cells during and after transformation. Exponentially growing secondary cultures of NIH mouse embryo fibroblasts were infected with 10^3 PFU/plate of Harvey strain of mouse sarcoma virus (H-MSV). At 2 and 5 days, during and after transformation of the cells respectively, the cultures were washed 5 times with Hank's minus glucose solution prewarmed to the assay temperature of 37 °C; 1.5 ml of prewarmed radioactive sugar containing 0.2 uC/ml with the indicated concentration of sugars in MEM without glucose, was then added, and incubation was carried out at 37 °C for 4 minutes. Cultures were then rapidly washed with 10 ml of buffered saline solution five times. The last wash was decanted and the remaining fluid removed under vacuum. Washed cells were scraped with a rubber policeman into 0.5 ml of water, transferred to a small tube, and homogenized briefly by vibration with a vortex mixer. Aliquots were determined for protein and radioactivity. The abscissa show 10^x M of extracellular sugar concentration with logarithmic scale and the ordinate shows the uptake rate by 10^y pmole/min/mg cell protein with logarithmic scale. A, B and C were assayed at 2 days post infection or during transformation; D, E and F were assayed at 5 days post infection or after completion of transformation. A and D represent the uptake of D-glucose, B and E for D-mannose, and C and F for D-galactose. The uptake of infected cells are shown by open circles, and that of control cells by closed circles. Reproduced with permission, from Hatanaka, M. (1976) J. Cell. Physiol 89, 745–750

Table 3. Transport of 2-deoxyglucose-[^{14}C] in MSV-infected cells. Cells at 5 days postinfection with MSV (Harvey) or postplating in the case of uninfected cells were assayed for sugar transport as described in the text. Incubation was carried out for 4 min at 37°. The figures given are average values of duplicate assays

| 2-deoxy-D-glucose | Uninfected | Infected |
M	nmoles/mg protein/min	
10^{-2}	15.6	83.2
10^{-3}	6.4	52.2
10^{-4}	0.95	9.7
10^{-5}	0.11	0.85
10^{-6}	0.01	0.11

The drastic change of uptake rates was observed at around 5×10^{-3}M sugar after transformation (Fig. 9). Thus, the transition of the sugar uptake system from a saturable to a nonsaturable process occurs near the physiological concentration of D-glucose normally seen in animal blood. At concentrations higher than 5×10^{-3}M, where a saturable process in barely involved, non-saturable transport of 2-deoxy-D-glucose proceeds ten to hundred-fold faster than the rate of simple diffusion of L-glucose (Table 3). These findings suggest that nonsaturable uptake of the sugars known to be substrates for the saturable transport carrier system may not be a physical process or simple diffusion as observed for L-glucose uptake. Rather, the nonsaturable uptake might be part of the physiological process which, along with the saturable process, is controlled by a membrane-coordination mechanism (Hatanaka 1976).

A plausible mechanism is a negative cooperativity of nutrient transport observed in bacteria (Anraku et al. 1973; Silharvy et al. 1974; Glover et al. 1975). Recently, Lipmann's group observed a similar nonsaturable process using the vesicle transport system derived from control and Rous sarcoma virus transformed chick embryo cells (Decker and Lipmann 1981).

Effects of Transported Sugars on Transformation and Tumorigenesis

Transformation of mouse fibroblasts by a murine sarcoma virus was influenced by the type of sugars transported from the culture medium. Exposure to 2-deoxy-D-glucose, 2-fluoro-2-deoxy-D-glucose and D-mannose caused a reduction in the number of transformed colonies *in vitro* and tumor formation *in vivo*. Table 4 and 5 show the summarized results. Although the precise nature of the inhibitory effect of the sugars on transformation and tumorigenesis remains to be elucidated, the data appear clear in showing that 2-deoxy-D-glucose, 2-fluoro-2-deoxy-D-glucose and D-mannose are effective substrates in the transport system of transforming cells. Several lines of evidence suggest that this effect is specific toxicity in cells. Selectivity was demonstrated by the fact that not all D-glucose analogs were inhibitory, e.g., halogenated sugars other than 2-fluoro-2-deoxy-D-glucose were marginally inhibitory at best. As shown

Table 4. Cell growth and focus formation in the presence of various sugars

Additions to medium	Cell growth		Focus formation	
	Without D-Glucose	With D-Glucose	Without D-Glucose	With D-Glucose
D-Glucose	+	+	+ + + +	+ + + +
D-Xylose	+	+	+ + + +	+ + + +
D-Galactose	+	+	+ +	+ +
L-Arabinose	Slow	+	+	+ +
D-Mannose	Slow	+	−	+
L-Xylose	Static	+	−	+ +
α-Methyl-D-glucose	Static	+	−	+ +
L-Mannose	−	+	−	+ + + +
L-Glucose	−	+	−	+ + + +
D-Inulin	−	+	−	+ + + +
D-Mannoheptulose	−	+	−	+ + + +
L-Fructose	−	+	−	+
β-D-Thioglucose	−	+	−	+
6-Deoxy-D-galactose	−	Toxic	−	−
2-Deoxy-D-glucose	−	Toxic	−	−
2-Fluoro-2-deoxy-D-glucose	N.D.	Toxic	N.D.	+
2-Dichloro-2-deoxy-D-glucose	N.D.	Toxic	N.D.	+ + +
3-Fluoro-3-deoxy-D-glucose	N.D.	+	N.D.	+ + + +
3-Fluoro-3-deoxy-D-xylose	N.D.	+	N.D.	+ + + +

Secondary cultures of fibroblasts from NIH Swiss mouse embryos (4×10^5 cells per plate) were cultured with 0.1% of additional sugar in Eagle's MEM with or without 0.1% D-glucose and 2% fetal-bovine serum. Cell growth (measured by counting cells per plate) was compared to that in Eagle's MEM with 0.1% D-glucose. "+" Means roughly similar growth rate as observed in Eagle's MEM; "Slow" means less than half the growth rate of the cells in Eagle's MEM; "static" means cells are intact but did not increase in number; "−" means most cells did not grow and eventually died; "toxic" means cell death with vacuolation. The cells cultured in the media indicated were inoculated with 0.1 ml 400 focus-forming units of MSV (Harver). 8 Days after inoculation, the foci were counted. "+ + + +" Indicates 100% of the cells were transformed; "+ +" indicates about 200 foci per plate; "+" indicates about 100 foci per plate; "−" indicates no foci at all. N.D. indicates not done

in Table 4, D-xylose supported the growth of cells with or without D-glucose, and was found to be ineffective in inhibiting transformation. Most importantly, conditions could be found in which cells grew well and still did not show signs of transformation (6×10^{-3}M D-mannose with 10^{-4}M D-glucose or 10^{-5}M 2-deoxy-D-glucose with 5×10^{-3}M D-glucose).

Perspectives and Exploitation

The availability of *in vitro* test systems to study D-glucose transport in cells made it possible for the first time to understand that the rate-limiting step of

Table 5. Reduced viral transformation in vivo by 2-deoxy-D-glucose and D-mannose

Sugar administered (μg)	No. of animals	Tumor incidence		
		17 day	20 day	37 day
0	15	14/15[a]	15/15	†
	26	11/26	25/26	†
		2-Deoxy-D-glucose		
1	6	6/6	6/6	†
10	10	4/10	9/10	†
100	21	3/21	8/21	10/21
	29	11/29	5/29	20/29
1000	15	0/6††	0/6	0/6
		D-Mannose		
100	13	3/13	4/13	11/13
1000	17	3/17	6/17	10/17

Newborn NIH Swiss mice were inoculated subcutaneously with 0.1 ml of 10^4 focus-forming units of MSV (Harvey), followed by 0.1 ml of sugar every day, until palpable tumors were observed. Most tumor-bearing mice died within 3–5 days after the formation of palpable tumors.
[a] Tumor-bearing mice to total number of mice.
†† Nine mice died 2 days after beginning the inoculation of high doses of 2-deoxy-D-glucose, probably due to the toxic effect

sugar uptake in cells is the transport across the plasma membrane. More than half a century ago, Warburg found that many tumor cells show enhanced aerobic and anaerobic glycolysis when compared to control (Warburg et al. 1930).

Consequently increased glycolysis is expected to lead to increased D-glucose uptake by tumor cells. However, the cells which have a rate limiting step at the transport may not necessarily result in increased D-glucose uptake. On the other hand, if the transport is enhanced by any mechanism, the initial uptake may increase whether or not the intra-cellular metabolism changes. We studied the possibility that the D-glucose transport was altered during *in vitro* cell transformation and consistently found that the transport system was enhanced, as evidenced by the initial rate of D-glucose uptake, during the process of transformation and continuing in the transformed state. To dissociate the increased D-glucose uptake from the overall change of glycolysis after transformation, 3-O-methyl-D-glucose and 2-deoxy-D-glucose were used. The kinetic analyses clearly demonstrate that the transport was enhanced by oncogenic transformation and that the transport was the first rate limiting step of D-glucose uptake in transformed cells. Based on these findings, the use of [^{11}C] 3-O-methyl-D-glucose and [^{18}F]-2-fluoro-2-deoxy-D-glucose may be most promising D-glucose analogs for detection of certain tumors by positron-emission tomography. This proposal, I believe, has the virtue of being immediately testable by radiologists who are interested in hopefully early and dynamic detection of neoplasms in man.

References

Anraku Y, Naraki T and Kanzaki S (1973) Transport of sugars and amino acids in bacteria VI. Change induced by valine in the branched chain amino acid transport systems of E. coli. J Biochem 73; 1149–1161

Glover GI, d'Ambrosio SM and Jensen RA (1975) Versatile properties of a nonsaturable, homogeneous transport system in bacillus subtilis: Genetic, kinetic, and affinity labeling studies. Proc Nat Acad Sci 72; 814–818

Decker S and Lipmann F (1981) Transport of D-glucose by membrane vesicles from normal and avian sarcoma virus-transformed chicken embryo fibroblasts. Proc Natl Acad Sci 78; 5358–5361

Graff JC, Hanson DJ and Hatanaka M (1973) Differences in cytochalasin B inhibition of 3-O-methylglucose uptake between Balb/3T3 cells and a murine sarcoma virus transformed clone. Int J Cancer 12; 602–612

Hatanaka M (1975) Transport of sugars across the tumor cell membrane. Cellular Membranes and Tumor Cell Behavior. Baltimore, Maryland, Williams and Wilkins Co, pp 114–172

Hatanaka M, Augl C and Gilden RV (1970a) Evidence for a functional change in the plasma membrane of murine sarcoma virus-infected mouse embryo cells. J Biol Chem 245; 714–717

Hatanaka M, Hanafusa H (1970b) Analysis of a functional change in membrane in the process of cell transformation by Rous sarcoma virus; Alteration in the characteristics of sugar transport. Virology 41; 647–652

Hatanaka M, Huebner RJ and Gilden RV (1969) Alterations in the characteristics of sugar uptake by mouse cells transformed by murine sarcoma viruses. J Natl Cancer Inst 43; 1091–1096

Hatanaka M (1974) Transport of sugars in tumor membranes. Biochim Biophys Acta (Review on Cancer) 355; 77–104

Hatanaka M (1982) Effect of D-glucose analogs on oncogenic transformation and tumors. In: Galeotti T et al (eds) Membranes in Tumor Growth. Amsterdam, Elsevier Biomedical Press, pp 375–379

Inui K, Moller DE, Tillotson LG and Isselbacher KJ (1979) Stereospecific hexose transport by membrane vesicles from mouse fibroblasts: Membrane vesicles retain increased hexose transport associated with viral transformation. Proc Natl Acad Sci, 76; 3972–3976

Kletzien RF and Perdue JF (1974) Sugar transport in chick embryo fibroblasts. I.A. Functional change in the plasma membrane associated with the rate of cell growth. J Biol Chem 249; 3366–3374

Kletzien RF and Perdue JF (1975) Induction of sugar transport in chick embryo fibroblasts by hexose starvation. J Biol Chem 250; 593–600

Lieb WR and Stein WD (1971) The molecular basis of simple diffusion within biological membranes. In: Bronner F, Kleinzeller A (eds) Current Topics in Membranes and Transport. New York, London, Academic Press, pp 1–39

Plagemann PGW and Richey DB (1974) Transport of nucleosides, nucleic acid bases, choline and glucose by animal cells in culture. Biochimica et Biophysica Acta 344; 263–305

Salter DW, Baldwin SA, Lienhard GE and Weber MJ (1982) Proteins antigenically related to the human erythrocyte glucose transporter in normal and Rous sarcoma virus-transformed chicken embryo fibroblasts. Proc Natl Acad Sci 79; 1540–1544

Silhavy TJ, Boos W and Kalckar HM (1974) The role of the Escherichia Coli galactose-binding protein in galactose transport and chemotaxis. In: Jeanicke L (ed) Biochemistry of Sensory Functions. Berlin Heidelberg New York, Springer, pp 25–27

Vyska K et al (1982) In this volume
Warburg O (1930) The metabolism of tumours (translated in English by Dickens, F)
    London, Constable
White MK, Bramwell ME and Harris H (1981) Hexose transport in hybrids between
    malignant and normal cells. Nature (Lond) 264; 232–235
Zylka JM and Plagemann PGW (1975) Purine and pyrimidine transport by cultured
    Novikoff cells. Specificities and mechanism of transport and relationship to phos-
    phoribosylation. J Biol Chem 250; 5756–5767

In Vivo Determination of Kinetic Parameters for Glucose Influx and Efflux by Means of 3-O-¹¹C-Methyl-D-Glucose, ¹⁸F-3-Deoxy-3-Fluoro-D-Glucose and Dynamic Positron Emission Tomography; Theory, Method and Normal Values

K. Vyska, M. Profant, F. Schuier, E. J. Knust, H.-J. Machulla,
H. M. Mehdorn, W. H. Knapp, G. Spohr, R. von Seggern,
I. Kimmling, V. Becker, and L. E. Feinendegen

Introduction

Imbalance between perfusion, transport and metabolism may determine the ultimate damage in ischemic brain disease (Mies et al. 1981, Pulsinelli et al. 1981). Therefore, for the quantitative assessment of ischemic brain disorders the knowledge of at least two parameters is necessary. One is local perfusion. The second parameter should relate to tissue metabolism, for example, to the glucose utilisation rate (Sokoloff et al. 1977, Phelps et al. 1979, Kuhl et al. 1980) or to the local unidirectional glucose transport rate (Vyska et al. 1980, Vyska et al. 1981, Vyska et al. 1982, Vyska et al. 1983, Kloster et al. 1981).

In the present study, for the determination of the unidirectional glucose transport rate, ¹¹C-methyl-D-glucose (CMG) and 3-¹⁸F-deoxyglucose (3 FDG) were used.

CMG (Kloster et al. 1981) is an analogue of glucose, which is transported across the blood brain barrier (BBB) by the same carrier as glucose (Betz et al. 1974, Bidder 1968, Cutler et al. 1971, Oldendorf 1976, Agnes et al. 1967, Betz et al. 1974, Buschiazzo et al. 1970, Pardridge et al. 1975, Czaki, 1956), yet it does not enter cellular metabolism (Cutler et al. 1971, Pardridge et al. 1975). It returns to the circulating blood. Also 3 FDG (Knust et al. 1982, Knust et al. 1983) is transported across the BBB by the same carrier as glucose, but it is slightly phosphorylated (Halama et al. 1983). In comparative studies between CMG and 3 FDG it could be demonstrated, that the amount of 3 FDG which is phosphorylated is so low that within the experimental error the 3 FDG can be used as an analogue of CMG for the measurement in vivo. The advantage of 3 FDG consists in the longer physical half-life of ¹⁸F when compared with ¹¹C.

Since the CMG and 3 FDG techniques, used in this study, are identical, the theoretical analysis presented in this paper is, for the sake of simplicity, related solely to the case when CMG is used as an indicator.

The CMG/3 FDG technique is based on the concomitant evaluation of time dependent changes of CMG/3 FDG blood and CMG/3 FDG tissue concentrations.

The present paper demonstrates that by means of the CMG/3 FDG technique the rate constants for glucose transport across the BBB may be determined in any selected brain area. It will be shown that these parameters permit the determination of the local unidirectional glucose transport rate (LUGTR); moreover, it is proposed to extend the method to measure simultaneously the local perfusion rate (LPR).

In the first part of the contribution, the theoretical aspects of the glucose transport are discussed, and an operational approach for application of the CMG/3 FDG technique is developed. In the second part, the values obtained in normal are discussed.

Theory

Capillary Flow and Glucose Exchange

The exchange of glucose between blood and tissue fluids is known to occur in capillaries. The circulation is in the capillaries divided into numerous small channels with a great broadening of the stream bed. As a result of this the rate of flow is slowed and the surface to which the blood is exposed is enormously increased.

The transport of glucose in capillaries is governed by two quite different mechanisms. First, there is a molecular diffusion as a result of concentration differences; second, glucose molecules are entrained by the moving blood and are transported with it.

The exchange of glucose between blood and tissue occurs on the capillary wall. Therefore, the capillary wall represents the reaction surface in the system under study[1]. Under steady state conditions, the number of molecules which pass the reaction surface per unit time equals the number of molecules arriving at the surface. This mass flux to the surface equals

$$j = -D\left(\frac{\partial c}{\partial r}\right) \text{ at reaction surface } r = r_0 \quad [\mu mol/min\ cm^2] \tag{1}$$

where $\left(\frac{\partial c}{\partial r}\right) r = r_0$ is the derivative along the outwart normal, evaluated at the reaction surface, and D the diffusion constant for glucose in plasma.

The transport of glucose across the reaction surface (blood brain barrier) was demonstrated to be a carrier facilitated transport and to behave as a first order enzyme catalyzed reversible reaction. Therefore, the number of molecules which pass the reaction surface at unit of time is given by (see Appendix I)

$$q = \frac{V_{max}}{K_M + c_{s_b} + c_{s_t}} \cdot (c_{s_b} - c_{s_t}) \quad [\mu mol/min\ cm^2] \tag{2}$$

where V_{max} [$\mu mol/min\ cm^2$] is the maximal reaction velocity, K_M the Michaelis-Menten constant [$\mu mol/cm^3$], c_{s_b} the glucose concentration at the reaction sur-

---

1 Since for our studies the properties of glucose transport (which are described by experimentally determined relationships) and not the spatial localization of transport enzymes in BBB are needed, the blood brain barrier was considered for the modelling proposes as being identical with a capillary surface

face at the blood side [$\mu mol/cm^3$], and c_{s_t} the glucose concentration at the reaction surface on the tissue side [$\mu mol/cm^3$]. Under steady state conditions, we have

$$j = q. \tag{3}$$

Substituting expressions (1) and (2) into (3), we obtain for glucose transport across the BBB

$$-D\left(\frac{\partial c}{\partial r}\right) r = r_0 = \frac{V_{max}}{k_M = c_{s_b} + c_{s_t}} \cdot (c_{s_b} - c_{s_t}) \text{ at the reaction surface.} \tag{4}$$

Under these conditions, if the capillary is considered to be a tube, the mass transfer of glucose in the capillary can be described by the following equation (see Appendix II)

$$\int_{S_i} c_i \cdot v \cdot (-n) dS - \int_{S_0} c_0 \cdot v \cdot n \, dS = 2\pi r_0 \int_0^{z_0} \frac{V_{max} \cdot (c_{s_b} - c_{s_t})}{K_M + c_{s_b} + c_{s_t}} dz \tag{5}$$

where r_0 is capillary radius, z_0 the capillary length, S_i the area of capillary at inflow, S_0 the area of capillary at the outflow crossection, c_i and c_0 are the glucose concentrations in plasma on the arterial and venous side of the capillary, respectively.

According to the integral mean value theorem, the eq. (5) can be written as follows:

$$\int_{S_i} c_i \cdot v \cdot (-n) dS - \int_{S_0} c_0 \cdot v \cdot n \cdot dS = 2\pi r_0 z_0 \cdot \frac{V_{max} \cdot (c_{s_b}(\xi) - c_{s_t}(\xi))}{K_M + c_{s_b}(\xi) + c_{s_t}(\xi)}$$
$$= \frac{2\pi r_0 \cdot z_0 \cdot V_{max} \cdot c_{s_b}(\xi)}{K_M + c_{s_b}(\xi) + c_{s_t}(\xi)} - \frac{2\pi r_0 \cdot z_0 \cdot V_{max} \cdot c_{s_t}(\xi)}{K_M + c_{s_b}(\xi) + c_{s_t}(\xi)} \tag{6}$$

where $c_{s_b}(\xi)$ and $c_{s_t}(\xi)$ are the values of glucose concentration at the reaction surface, at a special point ξ. This point, which is in general determined by a solution of the hydrodynamic equations in the capillary tube, is in our case not known, and, therefore, it must be determined experimentally.

For this, we considered the data of Betz et al. (1973), who studied the glucose influx rate in isolated dog brains under different conditions.

In the first study, Betz et al. determined the glucose extraction rate at a constant plasma flow rate per cm^3 of tissue, f_p [$ml/min \, cm^3$] but at different glucose arterial plasma concentrations $\overline{c}_i$ (tested range of $\overline{c}_i$ was 3–50 $\mu mol/cm^3$). In the second experiment, they determined the glucose influx rate at a constant glucose plasma concentration $\overline{c}_i$, but at different plasma flow rates per cm^3 of tissue (tested range of plasma flows was 0.20–0.80 $ml/min \, cm^3$).

Since these studies were related solely to the glucose influx, the second term in eq. (6), which describes the efflux, disappears. Hence we get:

$$\int_{S_i} c_i \cdot \mathbf{n} \cdot (-\mathbf{n}) \, dS - \int_{S_o} c_o \cdot \mathbf{n} \cdot \mathbf{n} \, dS = 2\pi r_o \cdot z_o \cdot \frac{V_{max} \cdot c_{s_b}(\xi)}{K_M + c_{s_b}(\xi) + c_{s_t}(\xi)} \qquad (7)$$

If in analogy to the studies of Sokoloff et al. 1977 and Lund-Andersen et al. 1976 the c_{s_t} is at first approximation neglected when compared to $k_M + c_{s_b}$, the eq. (7) can be written as follows:

$$\int_{S_i} c_i \cdot \mathbf{v} \cdot (-\mathbf{n}) \cdot dS - \int_{S_o} c_o \cdot \mathbf{v} \cdot \mathbf{n} \cdot dS = 2\pi r_o \cdot z_o \cdot \frac{V_{max}}{K_M + c_{s_b}(\xi)} \cdot c_{s_b}(\xi). \qquad (8)$$

Using this equation, the unidirectional extraction rate E (the value which was determined by Betz et al. 1973) can be calculated as follows:

$$E = \frac{\int_{S_i} c_i \cdot \mathbf{v} \cdot (-\mathbf{n}) dS - \int_{S_o} c_o \cdot \mathbf{v} \cdot \mathbf{n} \, dS}{\int_{S_i} c_i \cdot \mathbf{v} \cdot (-\mathbf{n}) dS} =$$

$$= \frac{1}{\int_{S_i} L \, c_i \cdot \mathbf{v} \cdot (-\mathbf{n}) \cdot dS} \cdot \frac{2\pi r_o z_o V_{max}}{k_M + c_{s_b}(\xi)} \cdot c_{s_b}(\xi) \qquad (9)$$

where $\int_{S_i} c_i \cdot \mathbf{v} \cdot (-\mathbf{n}) \cdot dS = \overline{c}_i \cdot \overline{v}_i \cdot S_i = \overline{c}_i \cdot f_c$ is the amount of glucose in the

plasma entering the capillary through the cross-section S_i per unit of time, $\overline{c}_i$ the mean glucose concentration in the cross-section S_i and f_c the plasma flow through the capillary cross-section S_i [ml/min].
This equation which can be modified as follows:

$$\frac{1}{E} = \frac{K_M}{n \cdot 2\pi r_o \cdot z_o \cdot V_{max}} \cdot \frac{n \cdot \overline{c}_i \cdot f_c}{c_{s_b}(\xi)} + \frac{1}{n \cdot 2\pi r_o \cdot z_o \cdot V_{max}} \cdot n \cdot \overline{c}_i \cdot f_c, \qquad (10)$$

where n is the number of capillaries per cm^3 of tissue [cm^{-3}] and $n \cdot f_c = f_p$ the plasma flow per cm^3 of tissue [ml/min cm^3], indicates that the value 1/E is a function of $n \cdot \overline{c}_i \cdot f_c = \overline{c}_i \cdot f_p$.
In order to study the dependence of 1/E on $\overline{c}_i \cdot f_p$, we plotted the data of Betz et al. 1973 in Fig. 1 as a function of $\overline{c}_i \cdot f_p$. In this figure, it can be seen that the dependence of 1/E on $\overline{c}_i \cdot f_p$ is, at least in the tested range of f_p and $\overline{c}_i$, linear. Therefore and since this straight line does not pass trough the origin, it becomes evident that eq. (10) is fulfilled if $c_{s_b}(\xi)$ is proportional to $\overline{c}_i \cdot f_p$, i.e.

$$c_{s_b}(\xi) = m \cdot f_p \cdot \overline{c}_i \quad [\mu mol/cm^3] \qquad (11)$$

where m is the proportionality factor having dimension [min].
Under these conditions, eq. (10) can be written as follows:

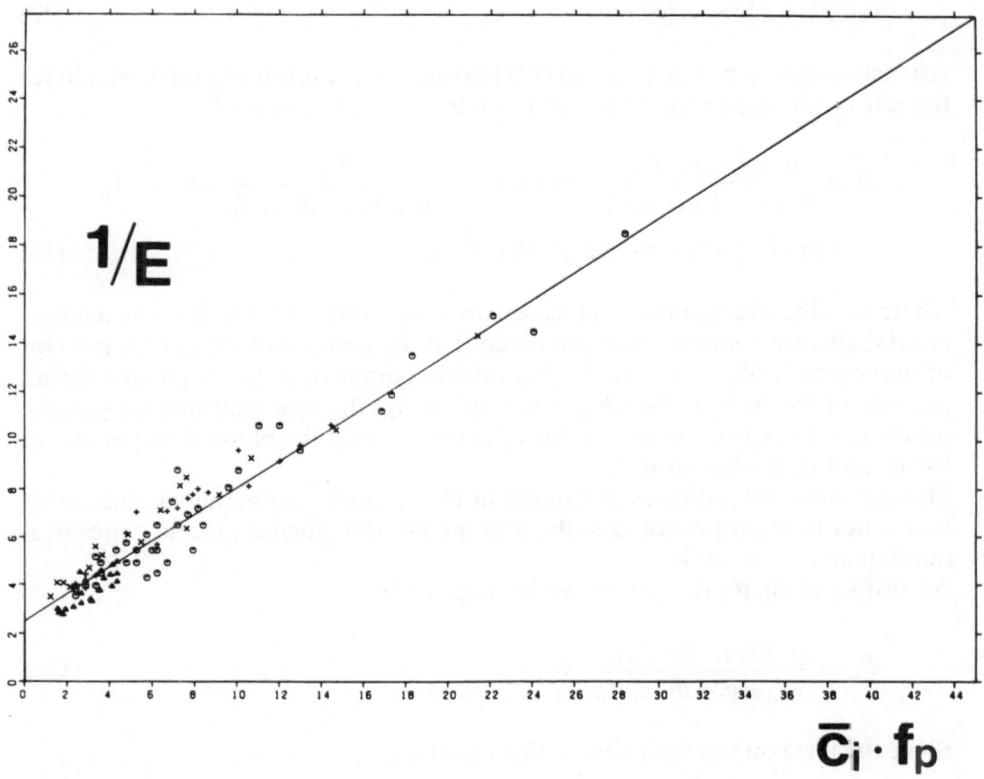

Fig. 1. $1/E$ as a function of $\overline{c}_i \cdot f_p \cdot (\Delta)$ data obtained by Betz et al. (1973) at a constant glucose plasma concentration, but at different plasma flow rates per cm^3 of tissue. ($\square \times o$) data obtained by Betz et al. (1973) at a constant plasma flow rate, but at different plasma concentrations

$$\frac{1}{E} = \frac{K_M}{n \cdot 2\pi r \cdot z_o \cdot V_{max} \cdot m} + \frac{1}{n\,2\pi r_o \cdot z_o \cdot V_{max}} \cdot n \cdot \overline{c}_i \cdot f_c. \tag{12}$$

The analysis of the data presented in Fig. 1 demonstrated, moreover, that

$$n\,2\pi r_o z_o \cdot V_{max} = 1,8 \,\mu mol/cm^3 \, min$$

and

$$K_M/m = 5 \,\mu mol/cm^3 \, min.$$

After taking into account that this value is in close agreement with data of Bachelard, 1974, who found the value of K_M to be $5\,\mu mol/cm^3$ we concluded that the value of m is close to 1 min.

Since, because of the counterflow orientation of capillaries, the changes of concentration along the capillary on the tissue side are minimal, the $c_{s_i}(z)$ was assumed to be a constant, i.e.

$$c_{s_t}(z) \sim c_{s_t}(\xi) \sim c_{s_t}(o) = c_t. \tag{13}$$

After substitution of eqs. (11) and (13) into eq. (7), the following relationship for the rate of glucose influx in cm^3 of brain tissue, Φ_i, is obtained:

$$\Phi_i = \frac{n \cdot 2\pi r_o \cdot z_o \cdot V_{max}}{K_M + c_t + m \cdot c_i \cdot f_p} \cdot m \cdot \overline{c}_i \cdot f_p = \frac{V_M}{K_M + c_t + m \cdot c_i \cdot f_p} \cdot m \cdot \overline{c}_i \cdot f_p$$

$$= m \cdot k_1 f_p \cdot \overline{c}_i = m \cdot k_1 \cdot (1\text{-}H_t) \cdot f \cdot \overline{c}_i, \tag{14}$$

where c_t is the average tissue glucose concentration [µmol/cm^3], $\overline{c}_i$ the average arterial glucose concentration [µmol/cm^3], n the number of capillaries per cm^3 of tissue [cm^{-3}], $V_M = n \cdot 2\pi r_o Z_o \cdot V_{max}$ the maximal velocity for glucose influx per cm^3 of tissue, $k_1 = V_M/(K_M + c_t + m \cdot \overline{c}_i \cdot f_p)$ the rate constant for glucose influx, $f_p = (1\text{-}H_t) \cdot f$, the plasma flow per cm^3 tissue, f the blood flow per cm^3 of tissue, and H_t the hematocrit.

The equation (14) indicates that under in vivo conditions the rate of glucose influx is not only proportional to the average arterial glucose plasma concentration but also to the flow.

According to eq. (6) the glucose efflux is given by

$$\Phi_e = \frac{n \cdot 2\pi \cdot r_o \cdot z_o \cdot V_{max}}{K_M + c_t + m \cdot \overline{c}_i \cdot f_p} \cdot c_t. \tag{15}$$

Since the term on the right side of this equation:

$$\frac{n \cdot 2\pi \cdot r_o \cdot z_o \cdot V_{max}}{k_M + c_t + m \cdot \overline{c}_i \cdot f_p} = \frac{V_M}{K_M + c_t + m \cdot \overline{c}_i \cdot f_p} = \tag{16}$$

is usually designated as the rate coefficient for glucose efflux, k_2 the eq. (15) can be written also as follows:

$$\Phi_e = k_2 \cdot c_t. \tag{17}$$

The eqs. (14) and (17) represents the basis for the interpretation of our experiments.

Model for CMG Kinetics

The CMG is a glucose analogue in which the hydrogen atom of the hydroxyl group at carbon-3 is replaced by ^{11}C-methyl group (Kloster et al. 1981) (Fig. 2).

CMG is transported across the blood brain barrier by the same carrier as glucose, but it is not phosphorylated or further metabolized (Betz et al. 1974, Czaky et al. 1956). In the model presented in Fig. 3 c_B^* [cpm/cm^3] is the concentration of CMG in blood; c_2^* [cpm/cm^3] is the CMG concentration in brain tissue;

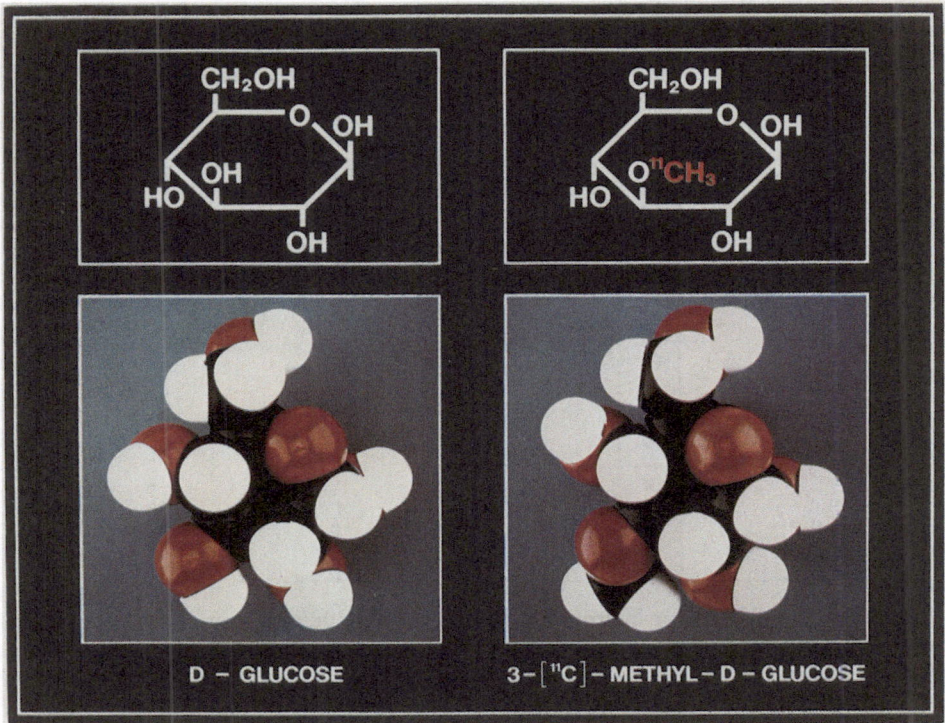

Fig. 2. Structure of 3-O-(C-11)methyl-D-Glucose

k_1^* [min^{-1}] is the rate constant for the CMG influx; k_2^* [min^{-1}] in the rate constant for CMG efflux.

In plasma, CMG competes with glucose for a common carrier for transport into a primary precursor pool in brain tissue. The rate of CMG accumulation in brain tissue (dc_2^*/dt) is equal to the difference between the rate of CMG influx (Φ_i) and CMG efflux (Φ_e):

$$\frac{dc_2^*}{dt} = \Phi_i^* - \Phi_e^* \tag{18}$$

As demonstrated above (see eq. (14)), the rate of glucose influx is proportional to the arterial plasma glucose concentration and to the flow. Consequently we assumed that also CMG influx is given by

$$\Phi_i^* = m \cdot k_1^* \cdot f \cdot (1\text{-}H_t) \cdot c_p^* \tag{19}$$

where f is the local perfusion rate, c_p^* local CMG plasma concentration, k_1^* rate constant for CMG influx and m proportionality factor having dimension min.

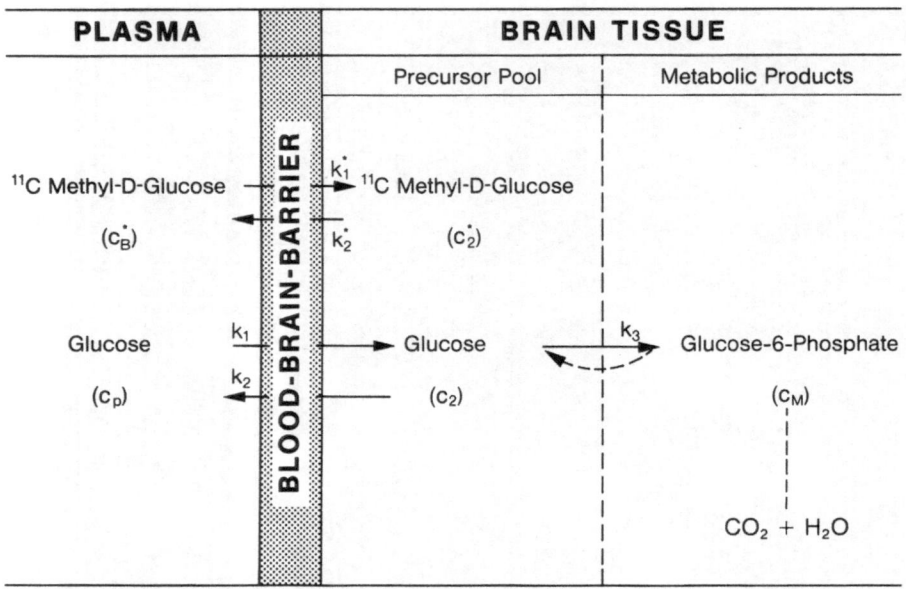

Fig. 3. The model used for interpretation of CMG measurements (modified according to Sokoloff et al. (1977)

Since

1. by external measurement the blood (c_B^*) and not the plasma (c_p^*) CMG concentration is registered and

2. according to Whitfield et al. 1974, the exchange of methylglucose between erythrocytes and plasma is so slow that it can be neglected in the first approximation, we assumed that plasma and blood CMG concentrations are related to each other by $c_B^* = (1-H_t) \cdot c_p^*$.

Under these conditions the eq. (19) can be written as follows:

$$\Phi_i^* = m \cdot k_1^* \cdot f \cdot c_B^*. \tag{20}$$

According to eq. (7) it can be assumed that the CMG efflux is given by

$$\Phi_e = k_2^* \cdot c_2^*. \tag{21}$$

Therefore

$$\frac{dc_2^*}{dt} = m \cdot k_1^* \cdot f \cdot c_B^* - k_2^* \cdot c_2^*. \tag{22}$$

As demonstrated in Appendix I, under the conditions of:

1. steady state of cerebral glucose utilization during the period of measurement,

2. a constant plasma glucose concentration,

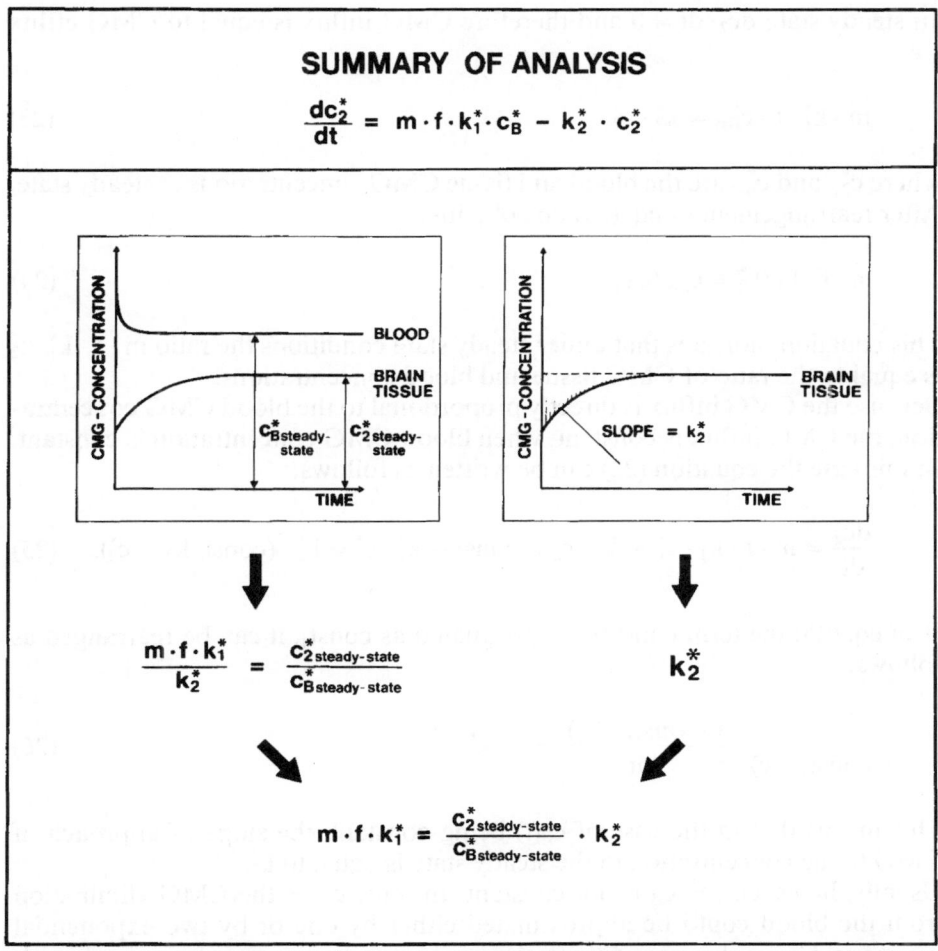

Fig. 4. Schematic representation of the evaluation of CMG data

3. symmetry of glucose transport across BBB and

4. application of CMG in tracer amounts,

the rate constant for CMG influx and efflux may be considered as true first order constants, which are independent of the plasma CMG concentration.
The elimination of the non-metabolizable CMG from the blood pool is so slow that it can be neglected and the mixing of the indicator in the blood is practically instantaneous so that the CMG blood concentration $c_B^*(t)$ can be considered to be constant. In this case the value of the ratio $m \cdot k_1^* \cdot f/k_2^*$ and the value of k_2^* can be determined, as demonstrated below, by the direct comparision of tissue and blood concentrations in steady state and by the analysis of the rate of approach CMG tissue concentration to steady state (see Fig. 4).

In steady state $dc_2^*/dt = 0$ and therefore CMG influx is equal to CMG efflux i.e.

$$m \cdot k_1^* \cdot f \cdot c_{Bss}^* = k_2^* \cdot c_{2ss}^* \tag{23}$$

where c_{2ss}^* and c_{Bss}^* are the blood and tissue CMG concentrations at steady state. After rearrangement of eq. (23) one obtains:

$$m \cdot f \cdot k_1^*/k_2^* = c_{2ss}^*/c_{Bss}^* \tag{24}$$

This equation indicates that under steady state conditions the ratio $m \cdot f \cdot k_1^*/k_2^*$ is equal to the ratio of CMG tissue and blood concentrations.

Because the CMG influx is directly proportional to the blood CMG concentration, the CMG influx is constant, when blood CMG concentration is constant. In this case the equation (22) can be written as follows:

$$\frac{dc_2^*}{dt} = m \cdot f \cdot k_1^* \cdot c_B^* - k_2^* \cdot c_2^* = const - k_2^* \cdot c_2^* = k_2^* \cdot (const/k_2^* - c_2^*). \tag{25}$$

If in eq. (25) the term $const/k_2^*$ is designated as $const_1$, it can be rearranged as follows:

$$\frac{1}{const_1 - c_2^*} \cdot \frac{d(const_1 - c_2^*)}{dt} = -k_2^*. \tag{26}$$

This means that in the case of $c_B^*(t)$ being constant, the slope of approach of CMG tissue concentration to the steady state is equal to k_2^*.

Usually, however, the c_B^* is not constant. In most cases the CMG elimination from the blood could be approximated either by one or by two exponential functions. Under these conditions, the parameters mentioned above can be calculated as described previously (Vyska et al. 1982).

Determination of Local Perfusion Rate f

As demonstrated above (see eq. (24)), the ratio of steady state CMG concentrations in tissue and blood is equal to the ratio $m \cdot f \cdot k_1^*/k_2^*$, i.e.

$$\frac{c_{2ss}^*}{c_{Bss}^*} = \frac{m \cdot f \cdot k_1^*}{k_2^*} \tag{27}$$

After rearranging of eq. (27), one obtains:

$$f = \frac{1}{m} \cdot \frac{k_2^*}{k_1^*} \cdot \frac{c_{2ss}^*}{c_{Bss}^*} \tag{28}$$

This relationship was used for the determination of f. Based on the discussion presented in Appendix I, the value of the ratio k_2^*/k_1^* was taken to be one. The value of m was determined in a separate study in five patients in which the local perfusion rate was determined in the brain by means of the ^{77}Kr technique just before the CMG study. Using results obtained in this study the value of m was determined on the basis of following formula:

$$m = \frac{1}{f_{kr}} \cdot \frac{k_2^*}{k_1^*} \cdot \frac{c_{2ss}^*}{c_{Bss}^*} \tag{29}$$

where f_{kr} is the value of perfusion determined by means of the ^{77}Kr technique. The results obtained in this study (see Fig. 8) indicate that the value of m is 1 min.

Determination of Local Glucose Influx Rate Φ_i

In order to determine the local unidirectional glucose transport rate Φ_i, the eq. (14) was considered:

$$\Phi_i = m \cdot k_1 \cdot f \cdot (1-H_t) \cdot c_p. \tag{30}$$

where c_p is the average plasma glucose concentration.
As demonstrated in Appendix I, the rate constants for glucose and CMG influx relate to each other as follows:

$$k_1 = k_1^* \cdot \xi \tag{31}$$

where ξ is the proportionality factor having value 1.1 (see Appendix I). Thus

$$\Phi_i = m \cdot \xi \cdot k_1^* \cdot f \cdot (1-H_t) \cdot c_p. \tag{32}$$

After substitution of eq. (27) in (32) one obtains:

$$\Phi_i = \xi \cdot \frac{c_{2ss}^*}{c_{Bss}^*} \cdot k_2^* \cdot (1-H_t) \cdot c_p. \tag{33}$$

This is the algorithm used in our studies for the determination of Φ_i.

Material and Methods

In the present studies, 2–5 mCi of CMG or 3 FDG were injected into an antecubital vein of the patient, and the transaxial activity distribution in one selected slice of brain was registered with the ECAT II scanner at 30 sec intervals for 40 minutes. Medium resolution shadow shields and high resolution data collection were used. The measured attenuation correction was applied for image reconstruction.

At different regions of brain scans ROIs were selected and the activity curves were plotted. Since the blood volume in brain is very small (cortex $\sim$ 4%, white matter $\sim$ 2%) the time activity curves registered over brain were considered as a measure for the brain CMG/3 FDG concentration, c_2^*.

As estimate for the capillary blood CMG/3 FDG concentration, c_B^*, either the arterial or venous CMG/3 FDG concentrations can be used. According to Sokoloff et al. 1977, the cerebral extraction ratio of the labeled glucose is very low (approximately 5%). This means that the mean capillary plasma concentration cannot differ from the arterial or venous plasma level by more than 5%. Therefore, it can be assumed that the arterial as well as venous blood concentrations are approximately equal to, or bear a constant relationship to, the capillary blood concentration (Sokoloff et al. 1977).

In the present study the activity registered over the superior longitudinal sinus (SLS) was taken as an estimate for the CMG/3 FDG venous concentration. This approach seems acceptable since by the use of the CMG/3 FDG technique the following conditions are fulfilled:

1. The non-metabolizable CMG/3 FDG is slowly eliminated from the blood pool, so that its concentration in blood remains relatively high during the whole measurement period; it is of the same magnitude as that in brain cortex.

2. according to our model, not the absolute CMG/3 FDG concentration in blood and in tissue, but their ratios are needed, and

3. from the spatial point of view, the SLS can be considered as being integrated in the body or the brain cortex.

Under these conditions and ROIs of SLS and brain cortex being of comparable size, the partial volume as well as crossover effects are nearly the same and they would practically cancel out in the ratio between the tissue and SLS activities. Consequently, the accuracy of the determination of the ratio tissue over blood CMG/3 FDG concentration would be higher than the accuracy of the determination of the absolute concentrations in the corresponding areas.

For the analysis of regional distribution of CMG/3 FDG in brain tissue the sum of the first 16 images registered in the dynamic study was used. For the analysis of time activity curves usually 6 ROIs were selected in cortex and 2 in white matter.

In order to test the applicability of the proposed method for the determination of the local perfusion rate in five cases, local perfusion rate was determined by the ^{77}Kr method. For this, krypton gas separated by a cold trap procedure was administered to the patient by inhalation from a closed system over a period of 4 min. (Yamamoto et al. 1983). The radioactivity distribution in a cross-section of the head was determined every 30 sec for a total period of about 20 min. Time-activity curves (clearance curves) were generated by the computer for each pixel (area: 1 $\times$ 1 cm) of the cross-sectional brain slice and the slopes of the semilogarithmically transformed curves were determined by a linear least squares fit. The calculated slope values F_i/λ_i were used to determine the blood

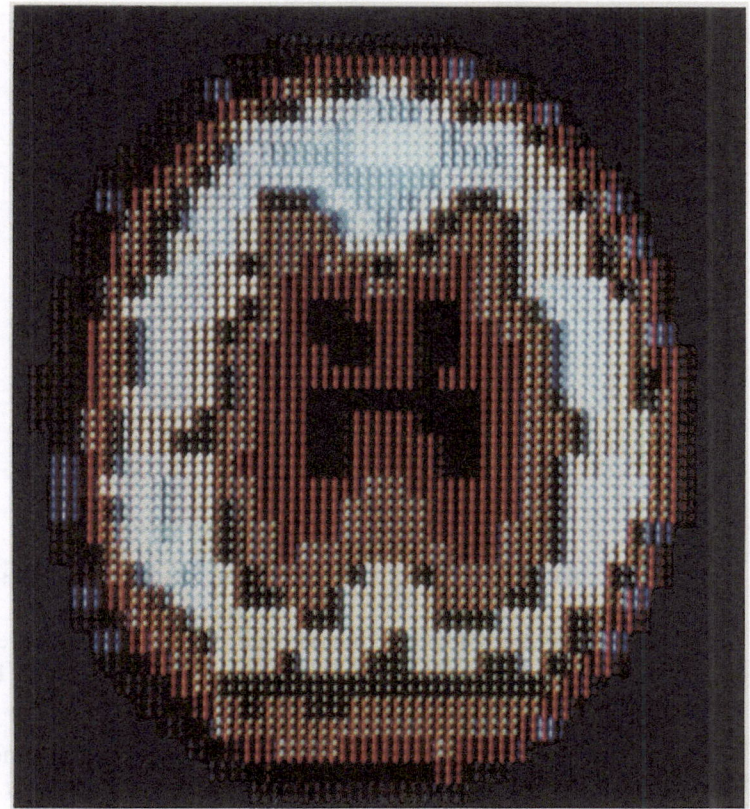

Fig. 5. CMG distribution pattern in normal brain

flow in ml/min per 100 g tissue. The tissue-blood partition coefficients of krypton for the tissue element were considered to be 0.94 for gray matter and 1.24 for white matter.

Results and Discussion

The CMG as well as 3 FDG were found to be effectively accumulated in normal brain cortex; significantly lower accumulation was observed in white matter (Fig. 5).

The time-activity curve registered over the ROI SLS (see Fig. 6) usually showed two exponnential components. The initial rapid decrease of activity probably reflects mixing of indicator in the blood pool and its equilibration with tissue. This was followed by slow indicator elimination ($T_{1/2} > 90$ min), indicating a high retention of non-metabolizable indicator in the blood.

The time-activity curves registered over different brain regions (Fig. 6) exhibited a rapid accumulation phase and a slow elimination phase. The CMG accumu-

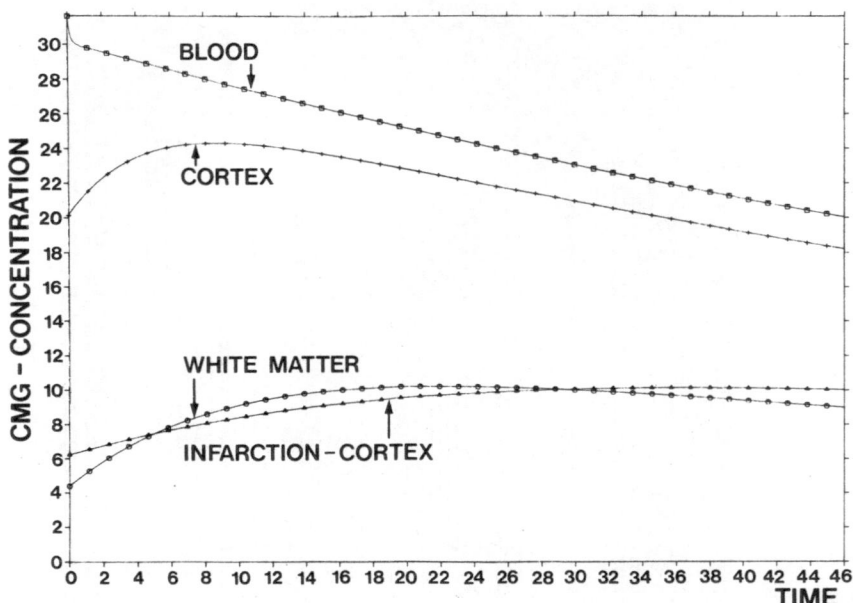

Fig. 6. Normalized, smoothed, and interpolated time activity curves registered over different brain regions after CMG injection (CMG concentration [cpm/pixel], time [min])

lation half times ranged from 1.8–3.2 min for normal cortex and from 5–7 min for normal white matter. These were significantly prolonged in ischemic areas. The values obtained from areas with destroyed blood brain barrier exceeded 10 minutes in cortex and 16 minutes in white matter.

At steady state distribution of the activity between tissue and blood, the CMG/3 FDG elimination from the brain tissue followed the time activity curve in blood.

Using the approach described above this time activity curve was used for the quantitative analysis of the data. The values of the ratio $m \cdot f \cdot k_1^*/k_2^*$ in normal cortex were determined to be 0.80–0.98 for both CMG and 3 FDG. In normal white matter values of 0.2–0.3 were observed. These values agree favourably with the data reported in the literature for the measurements in vivo (Sokoloff et al. 1977; Phelps et al. 1979; Reivich et al. 1982). In this connection it should be pointed out that in literature the value $m \cdot f \cdot k_1^* = \Phi_i^*/c_B^*$ is usually designated as rate constant for glucose influx.

The average value of k_2^* was determined to be for CMG 0.235 min^{-1} in normal cortex and 0.121 min^{-1} in normal white matter. These values are in good agreement with the data reported by Sokoloff et al. 1977 with ^{14}C-deoxyglucose and by Pardridge et al. 1975 with ^{14}C-methylglucose for rat brains and by Reivich et al. 1982 with ^{11}C-deoxyglucose for human brain. They are, however, higher than the value reported for humans by Phelps et al. 1979 with 2-^{18}F-deoxyglucose.

The value of k_2^* in cortex is approximately twice the value in white matter. This may be expected since the measured k_2^* is not only determined by the catalytic activity of the carrier system but also influenced by the total area of exchange surface, and it is known that the total capillary surface in the cortex is approximately twice as high as in white matter.

The local unidirectional glucose transport rate (LUGTR) was calculated to be 0.43–0.6 µmol/min g in normal cortex and 0.09–0.12 µmol/min g in white matter. These values are higher than the values reported with $2-^{18}F$-deoxyglucose (FDG) (Phelps et al. 1979; Phelps et al. 1982; Huang et al. 1980). This is expected, because CMG measures the glucose influx and not the net glucose transport rate, as does FDG.

The ratio of the glucose influx rate in cortex to that in white matter was approximately 4. This value is in good agreement with data reported by Kennedy et al. 1979.

The average value of k_2^* for 3 FDG was determined to be 0.47 ± 0.07 min^{-2}. This indicates a higher affinity of the glucose carrier system for 3 FDG than for CMG.

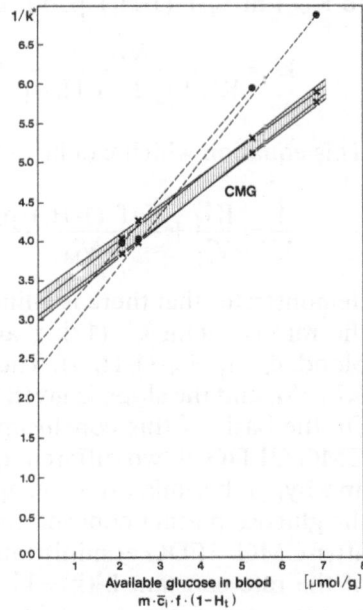

Fig. 7. The reciprocal values of the rate constant k^* at different glucose plasma concentrations.
(×——×) normal persons; (●---●) ischemic regions
These measurements were carried out with CMG

In this connection, it should be pointed out that in the case that the CMG/3FDG examination is carried out in the same patient at different glucose plasma concentrations, the data obtained provide the possibility of in-vivo determination of the maximal velocity (V_M) and Michaelis-Menten constant (K_M) for the glucose carrier system.

For this determination, it must be considered that according to eqs. (14), (16), (40) and (41) the following relationship is valid:

$$k_1^* = k_2^* = k^* = \frac{V_M^*}{K_M^* + c_t \cdot \rho + m \cdot c_i \cdot f_p \cdot \rho} = \frac{V_M^*}{K_M^* + c_t \cdot \rho + m \cdot c_i \cdot f \cdot (1-H_t) \cdot \rho}$$

where $k_1^* = k_2^* = k^*$ [min^{-1}] is the rate constant characterizing the activity of the carrier system for the CMG/3FDG transport across the BBB; V_M^* and K_M^* are the maximal velocity and Michaelis-Menten constants for CMG/3FDG transport; c_t is the glucose tissue- and $\overline{c_i}$ is the glucose plasma-concentration, f_p is the local plasma flow per gram of tissue, f is the local perfusion rate per gram of tissue, H_t is the hematocrit, m is the proportionality constant [1 min], and $\rho = K_M^*/K_M$.

If, in analogy to studies of Sokoloff et al. 1977 and Lund-Andersen et al. 1976, in this equation, $c_t \cdot \rho$ is – at a first approximation – neglected when compared to $K_M^* + m \cdot \overline{c_i} \cdot f \cdot (1-H_t) \cdot \rho$, one obtains:

$$k^* = \frac{V_M^*}{K_M^* + c_i \cdot f \cdot (1-H_t) \cdot \rho} \quad (m = 1 \text{ min})$$

This equation which can be modified as follows

$$\frac{1}{k^*} = \frac{K_M^*}{V_M^*} + \frac{\overline{c_i} \cdot f \cdot (1-H_t) \cdot \rho}{V_M^*}$$

demonstrates that there is a linear relationship between the reciprocal value of the rate constant k^*, $(1/k^*)$, and the amount of the available glucose in the blood, $\overline{c_i} \cdot f_p = \overline{c_i} \cdot (1-H_t) \cdot f$. The intercept of this line with the y-axis is equal to K_M^*/V_M^*, and the slope is given by ρ/V_M^*.

On the basis of this conclusion, we examined several patients with PET and CMG/3FDG at two different glucose plasma concentrations – normoglycemia and hyperglycemia after i.v. application of 10 g glucose (the determination of the glucose plasma concentration was performed just before and immediately after CMG/3FDG administration) – and determined the corresponding values of the rate constant $k_2^*(k_2^* = k_1^* = k^*)$ characterizing the catalytic activity of the glucose carrier system (see Material and Methods). The reciprocal value of the rate constant $(1/k^*)$ thus obtained is plotted in Fig. 7 as a function of the blood concentration of available glucose ($c_p \cdot (1-H_t) \cdot f$).

Using the relationship described above and assuming that the carrier facilitated transport of glucose is comparable to a unimolecular enzyme catalyzed reaction, i.e. that the changes of $1/k^*$ against the blood concentration of available glucose can be described by a straight line, the values of K_M^* and V_M^* were calculated from the data obtained, as follows:

1. the value of K_M^*/V_M^* was determined as the intercept of the straight line, connecting the $1/k^*$ values obtained in the same patient at low and higher glucose plasma concentrations.
2. the value of ρ/V_M^* is given by the slope of this straight line.

Assuming that according to Pardridge et al (1975) the value of ρ is 1.11 for CMG, the value of K_M^* was calculated to be 8.38 µmol/g and the value of V_M^* 2.61 µmol/min g in normal cortex.

In patients suffering from transient ischemic attacks, the value of K_M^* in the respective areas was reduced, indicating improved accessibility of the carrier system for glucose. The value of V_M^* was, however, significantly reduced, suggesting the destruction of some carrier units. In the case of a patient with diabetes mellitus, the value of K_M^* was significantly increased, demonstrating lower affinity of glucose to the carrier. In the same patient V_M^* was increased.

The local perfusion rate calculated according to equation (28) was 0.8–0.98 ml/min g in normal cortex and 0.2–0.3 ml/min g in white matter. These data are in good agreement with the data observed by Obrist et al. 1967 in normal humans by the use of ^{133}Xe-inhalation and blood sampling technique. They are slightly higher than the values reported by Ingwar et al. 1965, who injected ^{133}Xe into the internal carotid artery.

In order to test the validity of the CMG/3 FDG method for determination of local perfusion rate, 5 patients were also examined by [77]Kr technique. The results

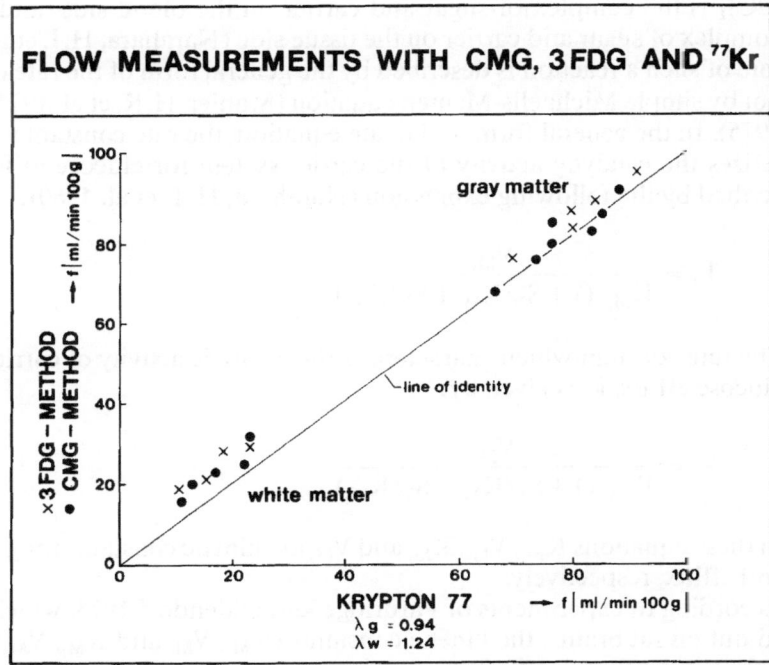

Fig. 8. Comparison of the values for the local perfusion rate, determined by CMG, 3 FDG and [77]Kr methods

obtained are summarised in Fig. 8. The diagram demonstrates a good correlation between both sets of data.

Summarizing the data presented so far we concluded that the CMG and 3 FDG in combination with dynamic positron emission tomography provide an excellent tool for the assessment of glucose transport across the BBB as well as for simultaneous detection of local perfusion rate.

Appendix I

The carrier-mediated transport of glucose across the BBB has been demonstrated by several investigators. As summarized by Pardridge and Oldendorf 1975, the blood brain barrier hexose carrier exhibits properties compatible with those of mobile carrier including saturable uptake, stereo-specificity, competitive inhibition with other hexoses and transport counterflow. Therefore, the glucose transport across the BBB is generally assumed to be comparible with a reversible unimolecular enzyme catalyzed reaction (Narahara, H.T. et al. 1960, Macey, R.I. 1979) i.e.

$$S_1 + C_1 \rightleftarrows (SC)_1 \rightleftarrows (SC)_2 \rightleftarrows C_2 + S_2, \tag{34}$$

where S_1 [μmol/cm^3] is the sugar in blood, S_2 [μmol/cm^3] is the sugar in tissue, C_1 is the carrier when it is on blood side, C_2 is the carrier on the tissue side, $(SC)_1$ is the complex of sugar and carrier on the blood side, and $(SC)_2$ is the complex of sugar and carrier on the tissue side (Narahara, H.I. et al. 1960). The rate of such a reaction is described by the general form of the rate equation and not by simple Michaelis-Menten equation (Mahler, H.R. et al. 1971; Segel, I.H. 1975). In the general form of the rate equation, the rate constant which characterizes the catalytic activity of the carrier system for glucose influx, k_1, is described by the following expression (Narahara, H.T. et al. 1960):

$$k_1 = \frac{V_{M_1}}{K_{M_1} \cdot (1 + S_1/K_{M_1} + S_2/K_{M_2})} \tag{35}$$

The rate constant which characterizes the catalytic activity of carrier system for glucose efflux, k_2 is given by:

$$k_2 = \frac{V_{M_2}}{K_{M_2} \cdot (1 + S_1/K_{M_1} + S_2/K_{M_2})} \tag{36}$$

In these equations K_{M_1}, V_{M_1}, K_{M_2} and V_{M_2} are kinetic constants for glucose influx and efflux, respectively.

According to experiments of Pardridge and Oldendorf 1975, which were carried out on rat brains, the kinetic constants (K_{M_1}, V_{M_1} and K_{M_2}, V_{M_2}) for glucose influx and efflux are the same, i.e.

$$K_{M_1} = K_{M_2} = K_M \text{ and } V_{M_1} = V_{M_2} = V_M.$$

Consequently, the equations (35) and (36) can be written as follows:

$$k_1 = k_2 = \frac{V_M}{K_M(1 + S_1/K_M + S_2/K_M)} \tag{37}$$

This means that the rate constant for glucose influx can be expected to be the same as that for glucose efflux.

This conclusion is in close agreement with the experimental data of Lund-Andersen et al. 1976, who found that the ^{14}C labelled 3-O-methyl glucose equilibrates in thick rat brain cortex slices at a concentration which relates to concentration of methyl glucose in the incubation medium in a ratio of 0.95 to 1.

Transport of CMG Across the BBB

Methyl glucose is transported across the blood brain barrier by the same carrier as glucose (Betz, A. L. et al. 1974; Czaky, T. Z. et al. 1956). In plasma CMG competes with glucose for a common carrier for transport through the BBB into brain tissue. Therefore the CMG influx into brain tissue, Φ_i^*, can be described by the general form of the rate equation, modified for the influence of the presence of the competitive substrate, glucose (Sokoloff et al. 1977):

$$\Phi_i^* = \frac{V_M^* \cdot S_1^*}{K_M^* \cdot (1 + S_1/K_M + S_2/K_M) + S_1^* + S_1^*} = k_1^* \cdot S_1^*. \tag{38}$$

In this equation Φ_i^* is the rate of CMG transport across the BBB measured under in vitro conditions, V_M^* and K_M^* the maximal velocity and Michaelis-Menten constant for CMG transport across the BBB, K_M the Michaelis-Menten constant for glucose transport across the BBB, S_1^* and S_1 the concentrations of CMG and glucose at blood side of reaction surface (BBB), S_2^* and S_2 the concentration of CMG and glucose at tissue side of reaction surface and k_1^* the rate constant reflecting the catalytic activity of the carrier system for CMG influx.

If the CMG is administered in tracer amounts, then

$$S_1 + S_2 \gg S_1^* + S_2^*$$

and

$$K_M^* \cdot (1 + S_1/K_M + S_2/K_M) \gg S_1^* + S_2^* \tag{39}$$

so that S_1^* and S_2^* become negligible quantities and

$$k_1^* = \frac{V_M^*}{K_M^* \cdot (1 + S_1/K_M + S_2/K_M)} \quad [min^{-1}] \tag{40}$$

Using an argumentation parallel to that described above, it can be demonstrated that

$$k_2^* = \frac{V_M^*}{K_M^*(1 + S_1/K_M + S_2/K_M)} \quad [\text{min}^{-1}] \tag{41}$$

This means that at

1. steady state of cerebral glucose utilisation during the period of measurement,

2. a constant plasma glucose concentration,

3. symmetry of glucose transport across BBB and

4. application of CMG in tracer amounts,

the rate constants for CMG influx and efflux are the same and may be considered as true first order constants which are independent of the plasma CMG concentration.
Dividing eq. (37) by eq. (40) leads to

$$\frac{k_1}{k_1^*} = \frac{V_M}{V_M^*} \frac{K_M^*}{K_M} = \xi. \tag{42}$$

Similarly the division of eq. (37) by eq. (41) leads to

$$\frac{k_2}{k_2^*} = \frac{V_M}{V_M^*} \cdot \frac{K_M^*}{K_M} = \xi. \tag{43}$$

If the values K_M^*, V_M^*, K_M and V_M determined by Pardridge et al. 1975 in rat brains are considered, the value of ξ is 1.11.

Appendix II

The Physical Model

Let us consider a substance with the density c, which is carried on by a incompressible fluid (blood).
The substance, which is transported by laminar flow of the blood diffuses across the pipe (see Fig. 9), so we obtain

$$\frac{\partial c}{\partial t} + \text{div}(cv) = \text{div}(D \, \text{grad} \, c). \tag{44}$$

At the wall we assume

$$- \mathbf{n} \cdot D \cdot \text{grad} \, c = f(c), \tag{45}$$

where $\mathbf{n}$ is the outward normal and D the diffusion constant.

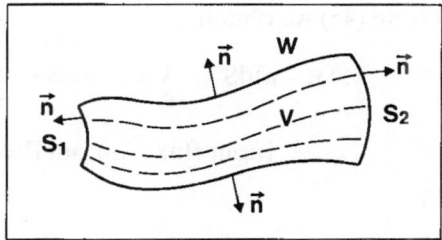

Fig. 9. The physical model used for the analysis of glucose transport into a brain tissue

Since in the case of steady state $\dfrac{\partial c}{\partial t} = 0$, the equation (44) can be written as follows:

$$\operatorname{div}(c\mathbf{v} - D\operatorname{grad}c) = 0. \tag{46}$$

This is an elliptical equation, which must be supported by boundary conditions (45) on the wall and by the boundary conditions at the entrance S_1 and exit S_2.
On the other hand

$$c\,\mathbf{v} - D\operatorname{grad}c = c\,\mathbf{v} - (D\operatorname{grad}c)_{\mathrm{II}} - (D\operatorname{grad}c)_{\perp} \tag{47}$$

where $(D\operatorname{grad}c)_{\mathrm{II}}$ is parallel to $\mathbf{v}$ and $(D\operatorname{grad}c)_{\perp}$ is perpendicular to $\mathbf{v}$.
The flux of fluid $\mathbf{v}$ is essentially larger than the diffusion flux $(D\operatorname{grad}c)_{\mathrm{II}}$ so that it can be neglected, and (46) can be written as follows:

$$\operatorname{div}(c\,\mathbf{v} - (D\operatorname{grad}c)_{\perp}) = 0. \tag{48}$$

This is a parabolic equation. In this case the boundary condition in exit S_2 is not required.
Applying the Gauss theorem on (48) we get

$$0 = \int_V \operatorname{div}(c\,\mathbf{v} - D\operatorname{grad}c)dV = \int_{S_1} (c\,\mathbf{v} - D\operatorname{grad}c)\mathbf{n}\,dS +$$

$$+ \int_{S_2} (c\,\mathbf{v} - D\operatorname{grad}c)\mathbf{n}\,dS + \int_W (c\,\mathbf{v} - D\operatorname{grad}c)\,\mathbf{n}\,dS \tag{49}$$

On the wall W is $\mathbf{v}\cdot\mathbf{n} = 0$ (for viscous flow $\mathbf{v} = 0$ at the wall; for non viscous flow $\mathbf{v} \perp \mathbf{n}$ at the wall).

At S_k is $\mathbf{n}\cdot(\operatorname{grad}c)_{\perp} = 0$ and $\left| \int_{S_k} c\,\mathbf{v}\,\mathbf{n}\,dS \right| \gg \left| \int_{S_k} (D\operatorname{grad}c)_{\mathrm{II}}\cdot\mathbf{n}\,dS. \right.$ $(k = 1, 2)$
So finally (49) implies

$$\int_{S_1} c\,\mathbf{v}(-\mathbf{n})dS - \int_{S_2} c\,\mathbf{v}\cdot\mathbf{n}\,dS - \int_W (-D\operatorname{grad}c)\mathbf{n}\,dS = 0. \tag{50}$$

Using (45) we obtain

$$\int_{S_1} c\, v(-\mathbf{n})dS \;-\; \int_{S_2} c\, v\cdot \mathbf{n}\, dS \;=\; \int_W f(c)dS. \tag{51}$$

Input flux   output flux   total losses

Summary

[11]C-methyl-D-glucose (CMG), 3-^{18}F-Deoxyglucose (3 FDG), and dynamic transaxial positron emission tomography (dPET) were used to measure the rate constants for glucose transport across the BBB in any selected brain area. The assay takes advantage of CMG and/or FDG being practically not metabolized in brain and being transported back from the tissue into the circulation; the simultaneous registration of tracer concentration in blood and tissue by dPET at 30 sec intervals for 40 min yields time activity curves which permit the recognition of tracer uptake by tissue to a steady state level and the CMG distribution between tissue and blood at the steady state.

The method was tested in 10 normal subjects. – In normal cortex the rate constant for CMG and 3 FDG efflux was at average 0.235 min^{-1} and 0.47 min^{-1} respectively. – The local glucose influx rate ranged from 0.43–0.6 µmol/min g in normal cortex and from 0.09–0.12 µmol/min g in normal white matter; local perfusion rate was calculated to be 0.8–0.98 ml/min g in normal cortex and 0.3–0.4 ml/min g in normal white matter.

References

1. Ackerman, RH, Carreia JA, Alpert NM Baron J-D, Gouliamos A, Grotta JC, Brownell GL, Taveras JM (1981) Positron imaging in ischemic stroke disease using compounds labeled with oxygen 15. Arch Neurol 38; 537–543
2. Agnew WF, Crone C (1967) Permeability of brain capillaries to hexoses and pentoses in the rabbit. Acta Physiol Scand 70; 168–175
3. Bachelard HS (1971) Specificity and kinetic properties of monosaccharide uptake into guinea pig cerebral cortex, in vitro. J Neurochem 18; 213–222
4. Betz LA, Gilboe DD, Yudilevich DL, Drewes L (1973) Kinetics of unidirectional glucose transport into the isolated dog brain. Am J Physiol 225; 586–592
5. Betz LA, Gilboe DD (1974) Kinetics of cerebral glucose transport in vivo. Inhibition by 3-O-methyl-glucose. Brain Res 65; 368–372
6. Betz AL, Gilboe DD, Drewes LB (1974) Effects of anoxia on net uptake and unidirectional transport of glucose into the isolated dog brain. Brain Res 67; 307–316
7. Bidder TG (1968) Hexose translation across the blood-brain interface: configurational aspects. J Neurochem 15; 867–874
8. Buschiazzo PM, Terrell EB, Regen DM (1970) Sugar transport across the blood brain barrier. Am J Physiol 219; 1505–1513
9. Cutler RWP, Sipe JC (1971) Mediated transport of glucose between blood and brain in the cat. Am J Physiol 120; 1182–1186
10. Czaky TZ, Wilson JE (1956) The fate of 3-O-[14]CH$_3$-glucose in the rat. Biochim Biophys Acta 22; 185–186

11. Halama JR (1983) Holden JE, Gatley SJ, Bernstein D, O'Hara KT, Ng CK, DeGrado TP (1983) Validation of F-18-3-Deoxy-3-Fluoro-D-Glucose (3 FDG) as an agent for measurement of glucose transport by positron emission tomography. J Nucl Med 24; P52

12. Heiss WD, Kloster G, Vyska K, Traupe C, Freundlieb C, Becker V, Feinendegen LE, Stöcklin G (1981) Regional cerebral distribution of ^{11}C-methyl-D-Glucose compared with CT perfusion patterns in stroke. J Cereb Blood Flow Metabol 1; Suppl 1: 506–507

13. Huang SC, Phelps ME, Hoffman EJ, Sideris K, Selin CJ, Kuhl DE (1980) Non-invasive determination of local cerebral metabolic rate of glucose in man. Am J Physiol 238; E69–E82

14. Ingwar DA, Cronquist S, Ekberg R, Risberg J, Hoedt-Rasmussen K (1965) Normal values of regional cerebral blood flow in man including flow and weight estimates of gray and white matter. Acta Neur Scand 41; Suppl 14: 72–84

15. Kennedy C, Sakurada O, Shinohara M, Jehle J, Sokoloff L (1979) Local cerebral glucose utilisation in the normal conscious macaque monkey. Ann Neurol 4; 293–301

16. Kloster G, Müller-Platz C, Laufer P (1981) 3-^{11}C-methyl-D-glucose a potential agent for regional cerebral glucose utilisation. Synthesis, chromatography, and tissue distribution in mice. J Lab Comp Radiopharm 18; 855–863

17. Knust EJ, Machulla H-J, Dutschka K (1982) Radiopharmaceuticals IV.: ^{18}F-Labelling with water target produced ^{18}F. Synthesis and quality control of ^{18}F-3-Deoxy-3-Fluoro-D-glucose. Radiochem Radioanal Letters 55; 1, 21–28

18. Knust EJ, Machulla H-J, Dutschka K, Molls M, Kafka Ch, Graebe K-J (1983) ^{18}F-3-Deoxy-3-Fluor-D-Glukose als potentieller Tracer für die Hirn- und Herzdiagnostik – Synthese und tierexperimentelle Untersuchungen. Nuc Compact 14; 40–44

19. Kuhl DE, Phelps ME, Kowell AP, Metter EJ, Selin C, Winter J (1980) Effects of stroke on local cerebral metabolism and perfusion: NH₃. Annals of Neurol 8; 47–60

20. Larsen OA, Lassen NA (1964) Cerebral hematocrit in normal man. J Appl Physiol 19; 571–574

21. Lund-Andersen H, Kjeldsen CS (1976) Kinetical analysis of the uptake of glucose analogs by rat brain cortex slices from normal and ischemic brain. In: Levi G, Battistin L and Lajtha A (eds) Transport phenomena in the nervous system: Physiological and pathological aspects. New York London. Plenum Press, pp 265–272

22. Mies G, Hossman KA (1981) Double tracer autoradiographic investigation of regional blood flow and glucose metabolism during spreading depression. J Cereb Blood Flow Metabol 1, Suppl. 1: 94–95, 1981

23. Narahara HT, Özand P, Cori CF (1960) Studies of tissue permeability VII. The effect of insulin on glucose generation and phosphorylation in frog muscle. J Biochem 235; 3370–3378

24. Macey RI (1979) Mathematical models of membrane transport processes. In: Andreoli TE, Hoffman JF and Fanestil DD (eds) Physiology of membrane disorders. New York London. Plenum Medical Book Company, pp 125–146

25. Mahler HR, Cordes EH (1971) Biological Chemistry 2nd Edition. New York. Evanston. San Francisco London. Harper and Row Publishers, pp 267–325

26. Obrist WD, Thompson HK, King CH, Wang HS (1967) Determination of regional cerebral blood flow by inhalation of ^{133}Xenon. Circ Res 20; 124–135

27. Oldendorf WH (1971) Brain uptake of radiolabeled amino acids, amines and hexoses after arterial injection. Am J Physiol 221; 1629–1639

28. Pardridge WM, Oldendorf WH (1975) Kinetics of blood brain barrier transport of hexoses. Biochim Biophys Acta 382; 377–382

29. Phelps ME, Huang SC, Hoffman EJ, Selin C, Sokoloff L, Kuhl DE (1979) Tomogra-

phic measurement of local cerebral glucose metabolic rate in humans with ^{18}F-Fluoro-2-deoxy-D-glucose: Valididation of method. Ann of Neurol 6; 371–388

30. Phelps ME, Mazziotta JC, Kuhl DE, Nuwer M, Packwood J, Metter J, Engel J Jr (1981) Tomographic mapping of human cerebral metabolism, visual stimulation and deprivation. Neurology 31; 517–529
31. Phelps ME, Mazziotta JC, Huang SC (1982) Study of cerebral function. J Cereb Blood Flow Metabol 2; 113–162
32. Pulsinelli W, Brierley J, Duffy T, Levy D, Plum F (1981) Ischemic neuronal damage, postischemic regional blood flow and glucose metabolism in rat brain. J Cereb Blood Flow Metabol 1; Suppl 1: 166–167
33. Reivich M, Greenberg J, Alavi A (1979) The use of fluorodeoxy-glucose technique for mapping of functional neural pathways in man. Acta Neurol Scand 60, Suppl 72; 198–199
34. Reivich M, Alavi A, Wolf A, Greenberg JH, Fowler J, Christman D, MacGregor R, Jones SC, London J, Shiue C, Yonekura Y (1982) Use of 2-deoxy-D(1-^{11}C)glucose for the determination of local cerebral glucose metabolism in humans: variation within and between subjects. J Cereb Blood Flow Metabol 2; 307–319
35. Segel IH (1975) Enzyme kinetics. New York London Sydney Toronto. Wiley – Interscience Publication, John Wiley, pp 34–39
36. Sokoloff L, Reivich M, Kennedy C, Des Rosiers MH, Patlak CS, Pettigrew KD, Sakurada O, Shinohara M (1977) The ^{14}C-deoxyglucose method for the measurement of local cerebral glucose utilisation: Theory, procedure, and normal values in the conscious and anesthetized albino rat. J Neurochem 28; 897–916
37. Vyska K, Höck A, Freundlieb C, Feinendegen LE, Kloster G, Stöcklin G (1980) 3-^{11}C-Methyl glucose a promising agent for in vivo assessment of function of myocardial cell membrane. J Nucl Med 21; P56–P57 (Abstr.)
38. Vyska K, Freundlieb C, Höck A, Becker V, Feinendegen LE, Kloster G, Stöcklin G, Traupe H, Heiss WD (1981) The assessment of glucose transport across the blood brain barrier in man by use of 3-^{11}C-methyl-D-Glucose. J Cereb Blood Flow Metabol 1, Suppl 1; 42–43
39. Vyska K, Freundlieb C, Höck A, Becker V, Schmid A, Feinendegen LE, Kloster G, Stöcklin G, Heiss WD (1982) Analysis of local perfusion rate and local glucose transport rate (LGTR) in brain and heart in man by means of C-11-Methyl-D-Glucose (CMG) and dynamic Positron Emission Tomography (dPET). In: Höfer R, Bergman H (eds) Radioaktive Isotope in Klinik und Forschung, 15. Band, Gasteiner Internationales Symposium 1982. Verlag H. Egermann, pp 129–142
40. Vyska K, Kloster G, Feinendegen LE, Heiss WD, Stöcklin G, Höck A, Freundlieb C, Aulich A, Schuier F, Thal HU, Becker V, Schmid A (1983) Regional Perfusion and Glucose Uptake Determination with ^{11}C-Methyl-Glucose and Dynamic Positron Emission Tomography. In: Heiss WD, Phelps ME (eds) Positron Emission Tomography of the Brain. Berlin Heidelberg New York. Springer, pp 169–180
41. Whitfield CF, Rames RS, Morgan HE (1974) Acceleration of sugar transport in avian erythrocytes by catecholamines. J Biol Chem 249; 4181–4188
42. Yamamoto YL, Meyer F, Menon D, Roland P, Diksic M (1983) Regional Cerebral Blood Flow Measurement and Dynamic Positron Emission Tomography. In: Heiss WD, Phelps ME (eds) Positron Emission Tomography of the Brain. Berlin Heidelberg New York. Springer

Physiological Conditions and Methodological Prerequisits for the In-Vivo Measurement of Substrate Transport in Tumors

W. H. Knapp, J. M. F. Chamayou, H. Ostertag, F. Helus, and S. Matzku

Introduction

During the last decades scintigraphy has emerged to play an essential role in tumor detection and staging. In view of the recent rapid development of newer imaging techniques, it appears useful to reflect on the clinical applicability of the specific potential of scintigraphy.

The essence of a tracer image may be defined as the sum of functional tissue properties. Thus, the image is able to mirror the pathomorphological or pathophysiological characteristics.

Classical tumor scintigraphy attempted to document pathomorphological changes to the organs involved in malignant disease. A particular physiological or metabolic parameter could not be assessed, however, because the accumulation of a radiotracer, following intravenous injection, depends on the sum of functional properties.

A particular difficulty of standard clinical nuclear medicine is due to the limited scope of radionuclides available, since they are confined to elements which do not have a physiological role in the examined organism, e.g. Tc-99m, Ga-67, In-111. Thus it is understandable that the most common procedures used for scintigraphic tumor detection (e.g. bone scintigraphy) have the same draw-backs as X-ray imaging: They do not permit quantification of tumor findings under therapy (Mc Killop et al. 1981; Goldstein et al. 1980).

There is an emphatic demand for the immediate documentation of tumor response to therapy, and for a more specific biological characterization of tumors for therapy control. The interest of oncologic nuclear medicine has recently focussed on labelled substrates utilized by rapidly growing tumors. A number of experimental studies sought to assess tumor-associated alterations in substrate utilization using either amino acids labeled with N-13 or C-11, or glucose derivates labeled with F-18 (Som et al. 1982; Hübner et al. 1980; Gelbard et al. 1979). In fact, clinical investigations demonstrated that e.g. N-13-L-glutamate provides adequate characterization of tumor response to chemotherapy in osteosarcoma (Rosen et al. 1979).

Considering the role of sugar and amino acid transport in malignancy (Hatanaka et al. 1974; Isselbacher 1972, Foster and Pardee 1969) would make it appear worth while to analyze whether the substrate uptake in tumors is governed by their increased transport capacity. Thus, the question remains to be answered which conditions have to be fulfilled in order to allow a quantitation of transport processes by in-vivo tracer studies.

Unidirectional Transport of Labeled Substrate

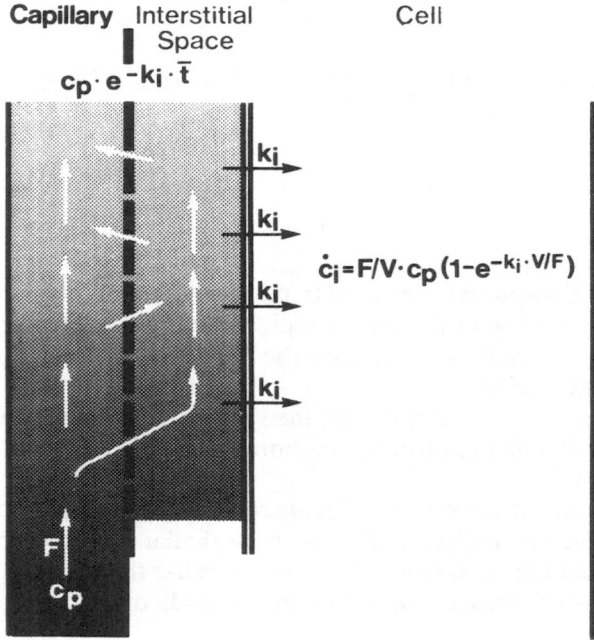

Fig. 1. Illustration of the model for the description of the substrate influx with symbols used in the text

Theory

In order to establish the relationship between the influx of activity from a given labeled substrate into tissue and the substrate transport by membranes, we have to define the rate-limiting transport process. According to Beck and Schultz (1972), glucose and amino acids are freely diffusible between the capillary system and the interstitial fluid space.

Using the definition of Bischoff (1975), we can therefore characterize the system under investigation as membrane limited.

Further the investigations on the diffusional and convectional behaviour of molecules in the interstitial fluid by Swabb et al (1974) showed that plasma and interstitial space can be considered as a continuous stirred tank reactor.

Under these circumstances, a simplification of the material balance equation in tumors is obtained by the reduction of the number of functional compartments. The unidirectional flux of the labeled substrate ($f\langle i\rangle$) into a given tissue volume can be equated as a function of arterial plasma concentration ($c\langle p\rangle$), plasma flow rate (F), the extracellular volume (V) and the transport rate coefficient $k\langle i\rangle$) – compare Fig. 1 –:

$$f\langle i\rangle(t) = F/V * c\langle p\rangle(t) * (1-\exp(-k\langle i\rangle * V/F)) \tag{1}$$

This formalism is an analogy to the model describing the loss of diffusible substances from a single capillary proposed by Renkin (1959) and Crone (1963). As in equation (1), it is assumed that labeled material which leaves the flow system, is promptly sequestered with none of it returning to the system, and that the rate of loss of material is proportional to the remaining concentration at each point.

Apart from the influx, we have to consider the efflux of the labeled substrate $(f\langle e\rangle)$ and of its metabolites $f\langle em\rangle)$ when considering the activity uptake $(A\langle c\rangle)$ within the intracellular compartment of a given tissue volume:

$$dA\langle c\rangle/dt = f\langle i\rangle - f\langle e\rangle - f\langle em\rangle \tag{2}$$

With equation (1) and

$$f\langle e\rangle = k\langle e\rangle * c\langle i\rangle(t) \tag{3}$$
$$f\langle em\rangle = k\langle em\rangle * c'\langle i\rangle(t) \tag{4}$$

we can substitute into equation (2), when $c\langle i\rangle$ and $c'\langle i\rangle$ are the intracellular activity concentrations from the labeled substrate and labeled metabolites, respectively, and $k\langle e\rangle$ and $k\langle em\rangle$ the rate coefficients for transport or diffusion from the intra- to the extracellular space.

If we assume that the unlabeled plasma substrate concentration is constant during the observation period, then the rate coefficients for the tracer become constants in analogy to the considerations by Sokoloff et al. (1977). We thus obtain

$$c\langle i\rangle(t) = k\langle i\rangle \left(\int_{t\langle o\rangle}^{t} c\langle p\rangle(z) * \exp((k\langle e\rangle + k\langle m\rangle)z)dz + a' * \exp-(k\langle e\rangle + k\langle m\rangle)t \right. \tag{5}$$

and

$$c'\langle i\rangle(t) = k\langle i\rangle k\langle m\rangle \left(\int_{T}^{t} \left(\int_{t\langle o\rangle}^{T} c\langle p\rangle(z) * \exp((k\langle e\rangle + k\langle m\rangle)z)\,dz \right. \right.$$
$$+ a' * \exp-(k\langle e\rangle + k\langle m\rangle)T * \exp(k\langle em\rangle T)dT + b' * \exp-k\langle em\rangle t \tag{6}$$

with $k\langle m\rangle$ representing a coefficient of the metabolic rate.

After substitution of equations (5) and (6) into the equations (3) and (4), and finally into equation (2), the differential equation can be solved for $A(t)$ with a closed form analytic expression. We therefore use the following formula to describe the input function:

$$c\langle p\rangle(t) = d(1-\exp-gt) + bt * \exp-ht \tag{7}$$

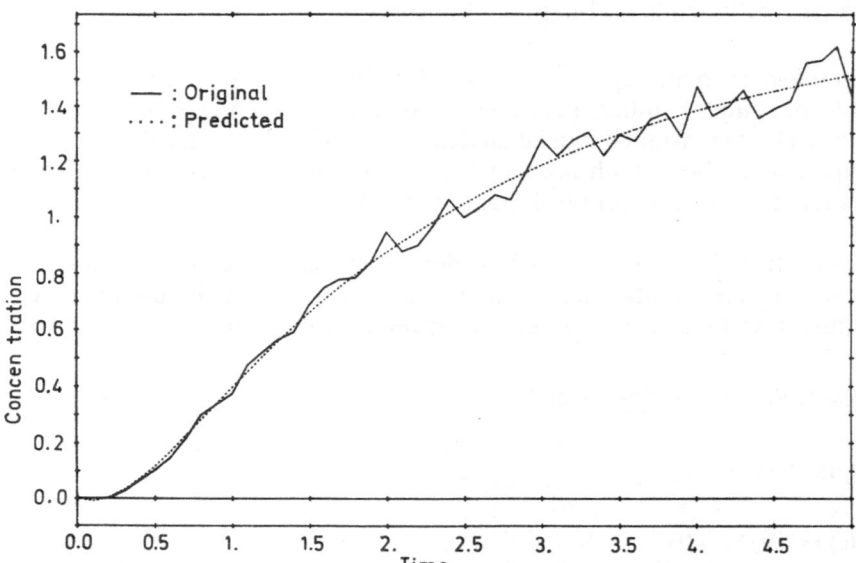

Fig. 2. Recalculation of the parameters $k\langle i\rangle$, $k\langle e\rangle$, $k\langle m\rangle$ after simulation of experimental errors (solid line, standard deviation = 10%) resulted in deviations of the same order of magnitude as the experimental error, when compared with the parameters given to the original function: $k\langle i\rangle$, $k\langle e\rangle$, $k\langle m\rangle > 0$; $k\langle em\rangle = 0$

The four parameters of this formula (b, d linear, g, h exponential) have to be fitted against the experimental curve obtained for $c\langle p\rangle(t)$.

Calculations

Using equation (7) we expressed A(t) as a function depending on time, on four known and four unknown parameters:

$$A(t) = H(k\langle i\rangle, k\langle e\rangle, k\langle m\rangle, k\langle em\rangle; b, d, g, h; t) \tag{8}$$

The parameters $k\langle i\rangle$ and $k\langle e\rangle$ are linear and the parameters $k\langle m\rangle$ and $k\langle em\rangle$ nonlinear (exponential or rational) in this function. The structure of this function can be used to advantage, if a separable non-linear least square routine is applied to fit the four parameters (Golub and Pereyra 1973; Bolstad 1977).

For certain numerical configurations, the non-linear parameter $k\langle em\rangle$ may cause instabilities, a fact which suggested the use of equation (7) and an analytical instead of a numerical resolution for the dynamic system (Chamayou 1979). Using this procedure, a symmetrical fitting structure is obtained for the four parameters of $c\langle p\rangle(t)$ and A(t).

Numerical computations were provided using the Monte-Carlo method for simulation of the experimental errors in the assessment of A(t), in order to recal-

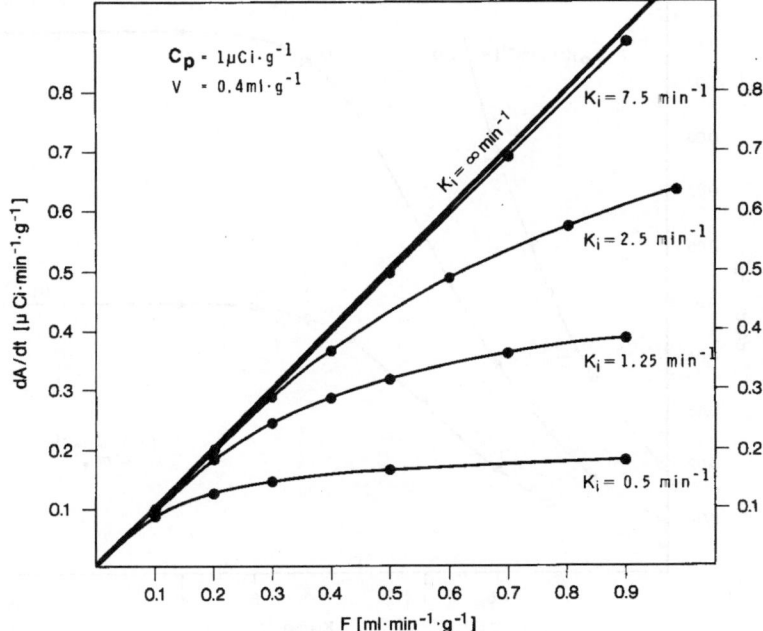

Fig. 3. Substrate influx as a function of plasma flow with different values of $k\langle i\rangle$

culate the already known parameters. The simulations showed that the function $A(t)$ must be determined with extreme accuracy (less than 1% error). This cannot be achieved with the experimental set-up. When this accuracy is not obtained, instabilities may occur leading to false fittings.

However, when the non-linear parameter $k\langle em\rangle$ is equal to zero (no release of labeled metabolites from the tissue), the results remain satisfactory even for large experimental errors exceeding 10%. An example is given in Fig. 2. For $k\langle m\rangle$, the only unknown, non-linear parameter in equation (8), the solution becomes unstable only when $k\langle m\rangle = h$.

Since $c'\langle i\rangle$ equals zero for $k\langle m\rangle = 0$ (no metabolism), no non-linear unknown occurs under non-metabolic conditions. Thus, the most complex physiological behaviour of a substrate which allows the equation (8) to be analyzed for $k\langle i\rangle$, is characterized by

$$k\langle i\rangle, \; k\langle e\rangle, \; k\langle m\rangle > 0 \text{ with } k\langle em\rangle = 0.$$

In practical terms, the following conditions have to be met when substrate transport is to be measured with tracers in vivo:
a) Unmetabolizable substrates or analogues have to be used or
b) Substrates or analogues have to be employed from which the label is not released from the intracellular space after metabolization or enzyme binding.
It is not only necessary to determine which degree of biologic complexity can still be resolved, it is also necessary to assess whether calculated values for $k\langle i\rangle$

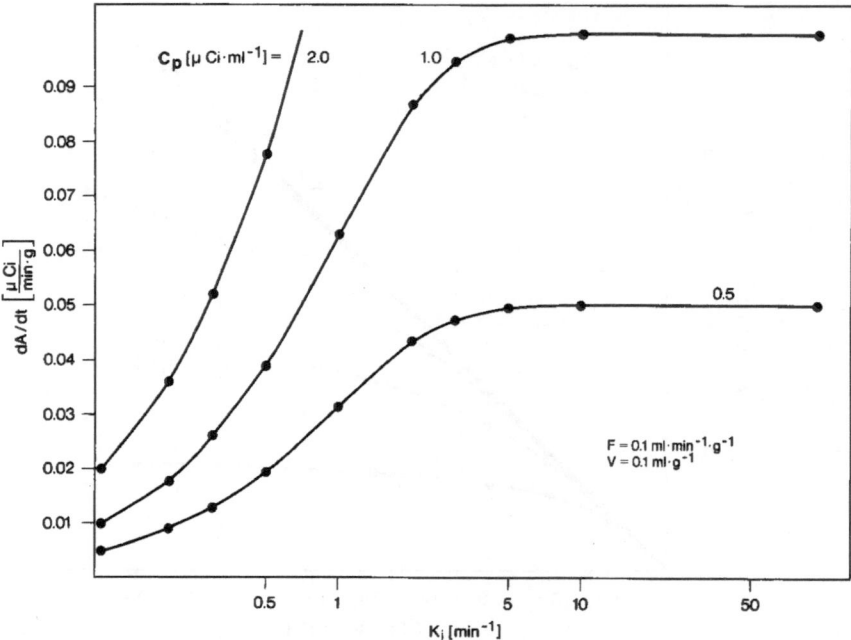

Fig. 4. Substrate influx as a function of $k\langle i\rangle$ at given flow

can be obtained with a sufficient degree of confidence. The range of confidence may become questionable for high values of $k\langle i\rangle$ and for low F/V ratios when random errors in the measurement of $A(t)$ are present.

Thus, the first term of equation (1): $f\langle i\rangle(t)$ expressed as dA/dt was plotted as a function of F while $k\langle i\rangle$ was varied (Fig. 3). Furthermore, the same first term was plotted as a function of $k\langle i\rangle$ during stable flow (Fig. 4). The critical range in which $k\langle i\rangle$ does not significantly influence $f\langle i\rangle(t)$, can be depicted.

If tumors are to be characterized by their transport rate coefficients, it is essential that the transport rates found can be experimentally separable from those of normal tissue, e. g. musculature. As shown in Fig. 3 and 4, low $k\langle i\rangle$ values for normal tissue are required when membrane properties are to be evaluated in vivo.

Experimental

N-13-L-glutamate was shown to be of value in characterizing the malignant activity of osteosarcoma and other neoplastic diseases (Rosen et al. 1979). The high tumor uptake of N-13 following the injection of the labeled substrate could suggest presence of increased amino-acid transport (Isselbacher 1972, Foster and Pardee 1969).

Using N-13-glutamate, we applied the presented model to analyze the experimental data, and evaluated

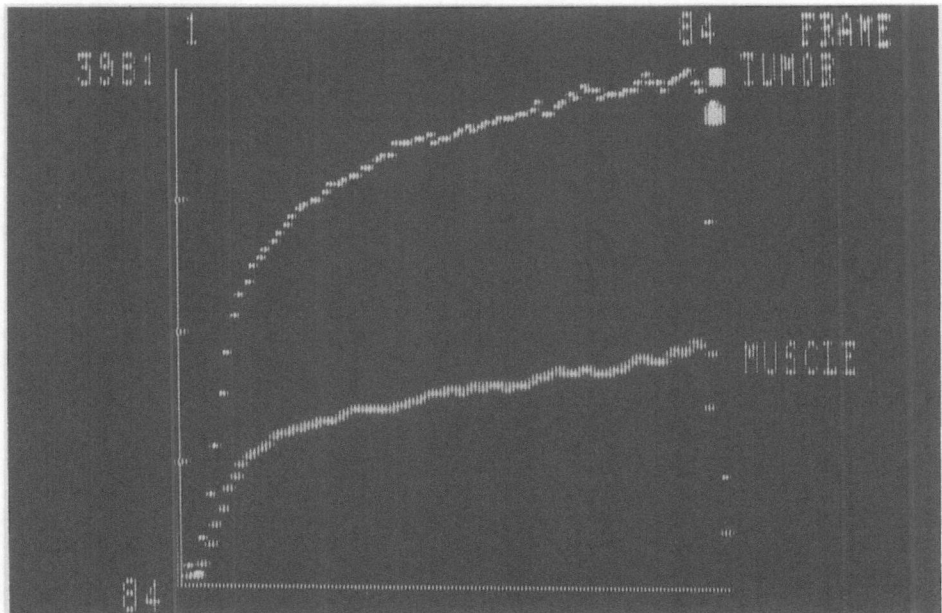

Fig. 5. Time-activity curve obtained from a human tumor (giant cell tumor = T) and from a control region (musculature = M), 0–14 min following i. v. injection of 5 mCi N-13-L-glutamate

(1) whether a stable solution for $k\langle i\rangle$ can be found and
(2) whether increased $k\langle i\rangle$ values, compared to normal tissue (musculature), produce a significant change of $A(t)$.

ad (1): The time-course uptake pattern of N-13-glutamate has been investigated in 8 human tumors (4 osteosarcoma, 1 hemangiopericytoma, 1 reticulosarcoma, 1 giant-cell sarcoma, 1 malign Schwannoma), in 25 transplant tumors in rats (5 BSp 27, 5 BSp 41, 5 BSp 73, 5 BSp 130, 5 BSp 141), and in 3 tumor transplants in rabbits (VX 2). The examination proceeded 10–20 min following systemic injection of N-13-L-glutamate. The preparation of the radiopharmaceutical and the imaging procedure have been described previously (Knapp et al. 1982). The analysis of the curves showed that in a majority of human tumors $A(t)$ was not significantly influenced by the wash-out term $f\langle em\rangle$ (Fig. 5).

Two human tumors and most of the tumor transplants in rats showed a different time-course which did not allow to neglect the washout term (Fig. 6). In the remaining 6 human tumors, $A(t)$ was approximated with $k\langle em\rangle = 0$. Thus it was shown that the kinetic behaviour of N-13-L-glutamate may allow a stable solution of the transfer equation of the radioactive label in tumors.

ad (2): In 3 rabbits, the term $1-\exp(-k\langle i\rangle * V/F)$ was determined by rapid sequential imaging (framing period = 0.1 sec) of the thigh. The examination followed

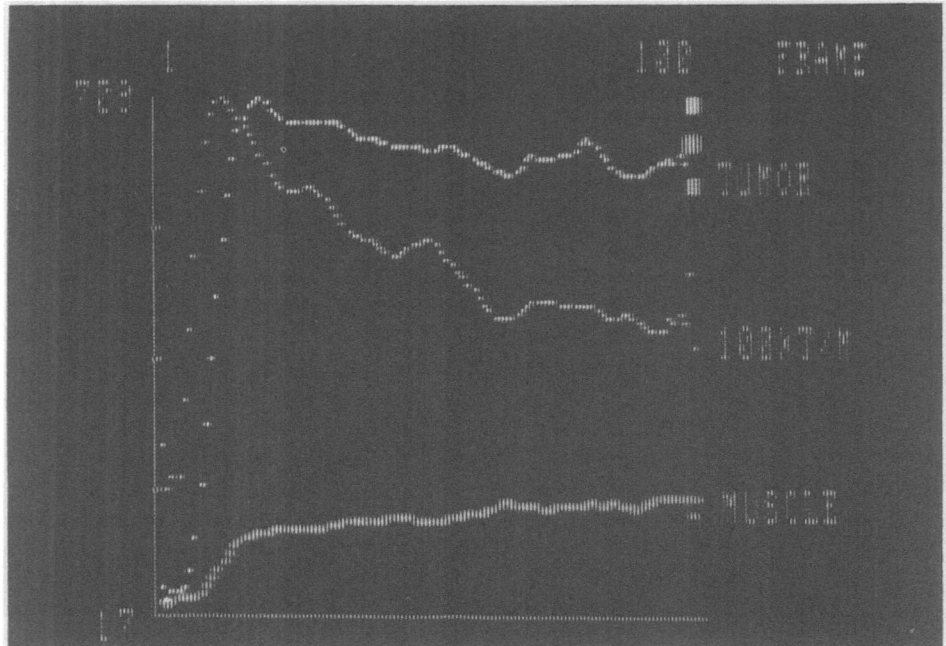

Fig. 6. Time-activity curve obtained from a transplanted tumor in the rat, 0–16 min following injection of 5 mCi N-13-L-glutamate. T = tumor, M = control region (musculature)

the bolus injection of 8 mCi N-13-L-glutamate into the a. femoralis. The single-pass residual fraction of activity of the thigh was determined, using an analogous procedure to one first proposed by Sejrsen (1970) and later developed and validated by Eichling et al (1974) and Raichle et al (1974). The residual fraction of activity was found to be 0.88–0.91.

Using representative values (e. g. 1/min) for specific flow (F/V) in muscle (Barcroft 1974) and a residual fraction of 0.90, the calculated $k\langle i\rangle$ value was 2.3/min. If we postulate that the measured quantity dA/dt (equation 2) has an error below 5%, then the calculated value for $k\langle i\rangle$ ranges between 1.9 and 3.0/min.

At identical flow rates, an increase in the transport-rate coefficients to any extremely high value, would result in an activity increase of less than 10% (compare Fig. 3).

Thus, it is obvious that under conditions of glutamate supply and transport found under normal circumstances, alterations of the transport rate coefficient result in negligible changes of intracellular activity. Within certain limits, it appears that intracellular activity can be approximated as a linear function of blood flow (compare the curve for $k\langle i\rangle = 2.5$ in Fig. 3). This was verified by a series of simultaneous N-13 glutamate uptake and perfusion measurements. For the perfusion studies we used J-121 microspheres. The preparation proce-

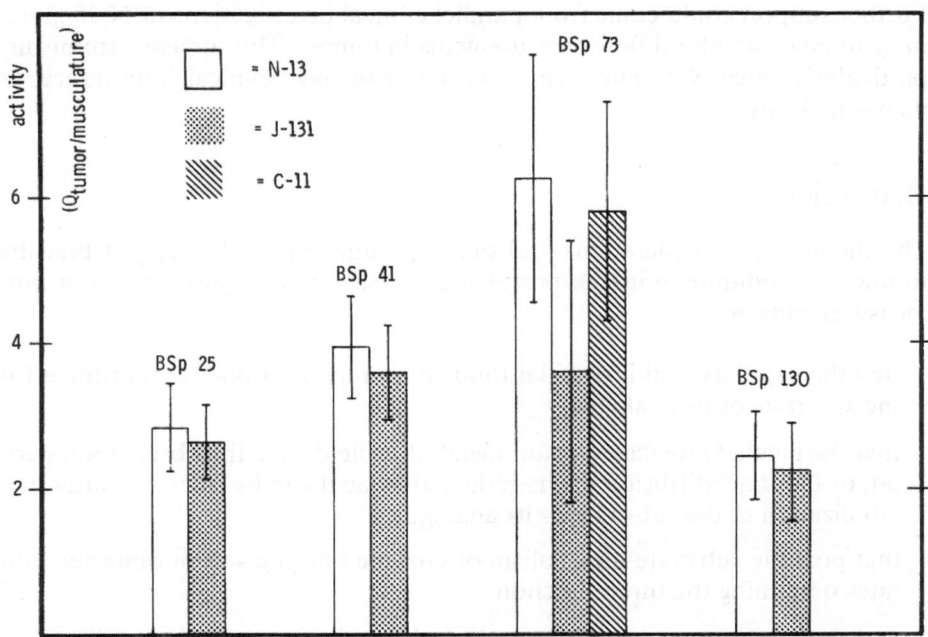

Fig. 7. Activity ratios tumor/musculature following injection of N-13-L-glutamate (2–5 min p.i.), J-121-microspheres and (in BSp 73) C-11-butanol (initial uptake). Significant lung activity in BSp 73 rats suggested a high degree of arterio-venous shunting of microspheres

dure has been described elsewhere (Knapp et al. 1983). The microspheres were injected into the aorta of rats. In the BSp 73 transplant tumor studied, we also used C-11 butanol (Oberdorfer et al 1982, Raichle et al 1976) for blood flow estimation. This appeared to be indicated, since arteriovenous shunting of the microspheres was suggested by a significant uptake of activity by the lung. The relative activity of N-13, I-121 and C-11 (in BSp 73) in four rat tumor types is demonstrated (Fig. 7). Relatively good agreement was observed between the N-13 accumulation and the uptake of either microspheres or C-11-butanol. These results support the postulated relationship between glutamate influx and perfusion.

The observed rate coefficient of glutamate influx into muscle suggested – as mentioned above – that the discrimination of even extremely high levels of $k(i)$ is not feasible with tracer techniques in vivo. The experimental data obtained from transplanted rat tumors in which these high levels might occur, support this standpoint, since uptake was predominantly related to flow. As literature does not give a clue to increased transport capacity being a general principle in all malignant tumors, increased levels of $k(i)$ cannot be postulated a priori for our tumors investigated. Therefore, the experimental data provide some evidence, but fail to provide absolute proof for the adequacy of our theoretical conclusions.

Further support could come from parallel clinical investigations of N-13 glutamate uptake and blood flow measurements in tumors. This appears promising, particularly since N-13-glutamate was found to have clinical importance, as mentioned earlier.

Conclusion

The theoretical considerations and our experimental results suggest that the following conditions must be postulated for substrate or analogue transport measurements in vivo:

– that the capillary and interstitial fluid space represent one compartment for the substrate or its analogue,
– that the labeled molecule be non-metabolizable during the observation period, or that backdiffusion of the radioactive label can be ruled out after metabolization of the substrate or its analogue,
– that possible substrate metabolism or enzyme binding will not interfer with rates describing the input function,
– that the transport rate coefficient which determines tracer influx, will not exceed a critical value at a given flow.

The methodological prerequisits include both the ability to measure intra- and extracellular space in a given tissue volume (e. g. using Br-75) and local blood flow (using $O-15-CO_2$ or C-11-butanol). These measurements can be carried out with positron-emission tomography. Furthermore, positron tomography permits quantitation of the tracer accumulation in tissue.
N-13-glutamate does apparently not provide tissue activity profiles which can be used to assess membrane transport. Future work will determine whether the following labeled substrates or analogues will be adequate for this purpose:

– F-18-2-deoxyglucose, F-18-3-deoxyglucose
– C-11-3-methylglucose
– C-11-aminocyclopentan-carboxylic acid
– C-11-alpha-C-methylated amino acids

Summary

The clinical need for in-vivo assessment of tumor progression and remission suggested the use of metabolic substrates labeled with positron emitters. The increased transport capacity of malignant cells for substrates has been established. Furthermore, determination of tumor uptake of N-13 labeled glutamate has been shown to be of clinical value.
We determined which conditions are required when membrane transport alterations in tumors are to be assessed in vivo. Furthermore, we sought to deter-

mine whether these conditions are present when N-13-glutamate is used as the tracer.

A model is presented which was used to calculate the time-course of intracellular activity as a function of arterial flow, transport rates of influx and efflux, and of the metabolic rate.

When experimental errors of 1–10% for each data point are postulated, the results show:

- that substrate analogues must be unmetabolizable, or that the label must not be released into the extracellular space following metabolization,
- that metabolic rates must not interfer with the input function,
- that for any given plasma flow per tissue volume, the transport-rate coefficient of normal tissue must not exceed a certain maximal value.

In the case of labeled glutamate, our data from animal experiments suggest that altered rate coefficients will only marginally influence uptake, when musculature is taken as reference. The principal factor responsible for glutamate uptake by tumors was found to be the perfusion.

A number of labeled compounds for positron-emission tomography are proposed which might be used for human tissue characterization based on evaluation of transport mechanisms.

References

1. Barcroft H (1974) Circulation in skeletal muscle. In: Hamilton WF, Dow PH (eds) Circulation Vol II. Baltimore. Williams & Wilkins, pp 1353–1386 (Handbook of Physiology)
2. Beck RE, Schultz JS (1972) Hindrance of solute diffusion within membranes as measured with microporous membranes of known pore geometry. Biochim Biophys Acta 255; 273–303
3. Bischoff KB (1975) Some fundamental considerations of the applications of pharmacokinetics to cancer chemotherapy. Cancer Chemother Rep 59; 777–793
4. Bolstad J (1977) Varpro Package. Computer Science Dpt. Stanford Univ.
5. Chamayou JMF (1979) Numerical experiments in identification of parameters in differential and partial differential equations. Comput Phys Comm 17; 217–226
6. Crone C (1963) Permeability of capillaries in various organs as determined by use of the indicator diffusion method. Acta Physiol Scand 58; 292–305
7. Eichling JO, Raichle ME, Grubb RL Jr., Ter-Pogossain MM (1974) Evidence of the limitations of water as a freshly diffusible tracer in brain in the rhesus monkey. Circulation Res 35; 358–364
8. Foster DO, Pardee AB (1969) Transport of amino acids by confluent and nonconfluent 3T3 and polyoma virus-transformed 3T3 cells growing on glass cover slips. J Biol Chem 244; 2675–81
9. Gelbard AS, Benua RS, Laughlin JS, Rosen G, Reiman RE, McDonald JM (1979) Quantitative scanning of osteogenic sarcoma with nitrogen-13 labeled L-glutamate. J Nucl Med 20; 782–784

10. Goldstein H, McNeil BJ, Zufall E, Treves S (1980) Is there still a place for bone scanning in Ewing's sarcoma? J Nucl Med 21: 10–12
11. Golub GH, Pereyra V (1973) The differentiation of pseudo-inverses and non-linear least squares problems whose variables separate. SIAM J Numer Anal 10; 413–432
12. Hatanaka M (1974) Transport of sugars in tumor membranes. Biochim Biophys Acta (Review on Cancer) 355: 77–104
13. Hübner KF, King P, Gibbs WD, Partain CL, Washburn LC, Hayes RL, Holloway E (1980) Clinical investigations with ^{11}C-labeled amino acids using positron emission computerized tomography in patients with neoplastic diseases. IAEA-SM-247/90
14. Isselbacher KJ (1972) Increased uptake of amino acids and 2-Deoxy-D-Glucose by virus-transformed cells in culture. Proc Natl Acad Sci US 69, 3; 585–589
15. Knapp WH, Helus F, Ostertag H, Tillmanns H, Kübler W (1982) Uptake and turnover of L-(^{13}N)-Glutamate in the normal human heart and in patients with coronary artery disease. Eur J Nucl Med 7; 211–215
16. Knapp WH, Helus F, Sinn HJ, Matzku S, Ostertag H, Brandeis WE, Braun A (1983) ^{13}N L-glutamate uptake in malignancy. Submitted for publication
17. McKillop JH, Etcubanas E, Goris ML (1981) The indications for and limitations of bone scintigraphy in osteogenic sarcoma: a review of 55 patients. Cancer 48: 1133–1138
18. Oberdorfer F, Helus F, Maier-Borst W, Silvester DJ (1982) The synthesis of (1-^{11}C)-butanol. Radiochem Radioanal Letters 53, 4; 237–252
19. Raichle ME, Eichling JO, Grubb RL Jr (1974) Brain permeability of water. Arch Neurol 30; 319–321
20. Raichle ME, Eichling JO, Straatmann MG, Welch MJ, Larson KB, Ter-Pogossian MM (1976) Blood-brain barrier permeability of ^{11}C-labeled alcohols and ^{15}O-labeled water. Am J Physiol 2; 543–552
21. Renkin EM (1959) Transport of potassium-42 from blood to tissue in isolated mammalian skeletal muscles. Am J Physiol 197; 1205–1210
22. Rosen G, Gelbard AS, Benua RS, Laughlin J, Reiman RE, McDonald JM (1979) N-13 glutamate scanning to detect the early response of bone tumors to chemotherapy. Proc Am Assoc Cancer Res., March 1979; 189
23. Sejrsen P (1970) Single injection, external registration method for measurement of capillary extraction. In: Crone C, Lassen N (eds) Capillary Permeability, Academic, New York, pp. 256–260
24. Sokoloff L, Reivich M, Kennedy C, DesRosiers MH, Patlak CS, Pettigrew KD, Sakurada D, Shinohara M (1977) The (^{14}C)Deoxyglucose method for the measurement of local cerebral glucose utilization: Theory, procedure, and normal values in the conscious and anesthetized albino rat. J Neurochem 28; 897–916
25. Som P, Atkins HL, Bandoypadhyay D, Fowler JS, Mac Gregor RR, Matsui K, Oster ZH, Sacker DF, Shiue CY, Turner H, Wan C-N, Wolf AP, Zabinski SV (1980) A fluorinated glucose analog, 2-Fluoro-2-deoxy-D-glucose (F-18): Nontoxic tracer for rapid tumor detection. J Nucl Med 21; 670–675
26. Swabb EA, Wei J, Guillino PM (1974) Diffusion and convection in normal and neoplastic tissues. Cancer Res 34; 2814–2822

Diagnostic Demands in Clinical and Experimental Oncology: Application of Substrates Labeled with Positron-Emitting Radionuclides

R. E. Reiman, G. Rosen, A. S. Gelbard, R. S. Benua, S. D. J. Yeh, and J. S. Laughlin

1. Introduction

In recent years, new regimens for the treatment of a variety of malignant tumors have led to greatly increased disease-free survival rates. For example, the introduction of high-dose methotrexate (HDMTX) with citrovorum factor rescue by Jaffe (1) for osteogenic sarcoma was a major breakthrough in the treatment of that highly malignant bone tumor. Later, Jaffe et al. (2), Rosen et al. (3) and Sutow et al. (4) found that primary osteogenic sarcoma, and its systemic metastases, could be better controlled by using HDMTX in combination with other drugs such as doxorubicin (adriamycin) and cyclophosphamide (cytoxan). In 1976, Rosen and co-workers reported their results in a series of patients who were treated with HDMTX and adriamycin preoperatively in an attempt to control and shrink the primary tumor and to eradicate metastatic disease at an early point in the regimen. These patients were noted to have a better disease-free survival than patients treated with post-operative adjuvant chemotherapy (5). Recent regimens for osteogenic sarcoma and Ewing sarcoma have further increased the survival of patients with these diseases (6, 7, 8). Today, more than 80% of patients with osteogenic sarcoma can expect to survive 5 years following intensive chemotherapy (9), compared to less than 20% in 1970 (10).

These regimes utilizing preoperative chemotherapy for the primary tumor place a greater burden on diagnostic technology. The clinical scope of treating the disease is no longer limited to the destruction of systemic metastases but is extended to the control of the primary tumor prior to ablative surgery or other local treatment. Hence, it is necessary to be able to document primary tumor regression or indeed progression, before the latter becomes clinically evident, so that the course of treatment can be modified or definitive local therapy can be carried out while still possible. As has been pointed out (6), each patient may have a different response to the treatment protocol, and hence it is of paramount importance that their progress be carefully evaluated, from the standpoint of drug toxicity as well as tumor response:

> "... it is advantageous to find for each individual patient the dose of HDMTX with CFR which elicits a response in the primary tumor, i.e., the effective dose for that patient ... The need to continuously and carefully evaluate (by all parameters available) the effect of chemotherapy on the primary tumor cannot be stressed strongly enough."

A number of clinical indicators can be used in certain cases to evaluate the effects of therapy upon primary and metastatic malignant disease. Biochemical markers, such as elevated alkaline phosphatase in osteogenic sarcoma, or elevated excretion of catecholamines in the urine in neuroblastoma, have been found useful in determining early tumor response. If the pre-therapy level of one of these markers is not elevated (as is the case for approximately 50 percent of patients), then they are not useful for this purpose. Radiographic changes in bone can be correlated with histologic response for patients with osteogenic sarcoma (11), as shown recently by Smith and co-workers. However, these changes usually take up to three months to become evident. Such techniques may be less useful for tumors limited to the soft tissues. Nuclear imaging modalities, such as scanning with Ga-67 gallium citrate (12) or Tc-99m labeled polyphosphates (13, 14), have proven useful in the detection of early primary or metastatic disease that is radiographically and clinically occult. The non-physiologic nature of these agents, and the resulting uncertainties in their mechanism of localization, have not made them attractive to the experimental oncologist, who would prefer to have a more physiological radioindicator of tumor response. Unfortunately, physiologic radiolabels such as N-13 and C-11 have short half-lives and must be produced at the site of use, and, until recently, scanning equipment which was capable of forming images with the high-energy positron-annihilation gamma rays was not available.

In the early 1970's, instrumentation was designed at a number of laboratories for producing quantitative, transaxial images *in vivo* of the distribution of compounds labeled with positron-emitting radionuclides. In turn, this stimulated much interest in the development of labeled physiologic substrates, such as amino acids, for use with the new modality known as positron-emission tomography (PET). The first reported enzymatic syntheses of N-13 labeled L-glutamine and L-glutamate (15, 16) were performed using N-13 ammonia produced by deuteron bombardment of methane. However, this method yielded N-13 ammonia in low specific activity with labeled and unlabeled impurities, which had to be removed by distillation. To improve the yields, specific activity and purity, investigators at the Sloan-Kettering Institute bombarded a water target with protons to produce N-13 labeled nitrates and nitrites, which were reduced to N-13 ammonia with Devarda's alloy in basic solution. The ammonia was subsequently used to synthesize N-13 L-glutamine and L-glutamate (17) in the multi-millicurie amounts necessary for imaging with a High Energy Gamma (HEG) scanning system (18). Gelbard, McDonald and co-workers demonstrated increased uptake of N-13 glutamine and glutamate (19, 20) in spontaneous tumors in dogs, and later in human osteogenic sarcoma (21, 22, 23), Ewing sarcoma (24), embryonal rhabdomyosarcoma (25) and tumors of neurogenic origin (26). The acquisition of a prototype total-body positron-emission tomograph (PC-4200, The Cyclotron Corp., Berkeley, CA, USA) in 1978, and subsequent three-dimensional imaging studies with F-18 sodium fluoride (27) and C-11 carbon dioxide (28), demonstrated the feasibility of evaluating tumor response in patients using this instrument.

In this work we summarize our experience to date with N-13 labeled glutamate as a tool in the evaluation of the response of malignant disease to chemotherapy. We will review how changes in N-13 tumor uptake during chemotherapy correlate with clinical and histological indicators of response, demonstrate the potential of N-13 labeled amino acids as agents for use in positron-emission tomography, and discuss the significance of these quantitative, *in vivo* studies to the clinical and experimental oncologist.

2. N-13 L-Glutamate: Two-Dimensional Imaging Studies

Since 1978 we have performed over 250 imaging studies in patients with a variety of malignant tumors, in normal volunteers (29) and in patients with myocardial hypertrophy secondary to valvular dysfunction (30). The results of a number of tumor studies are documented elsewhere (24, 25, 26). We will review the findings on a series of patients with osteogenic sarcoma.

N-13 glutamate was prepared by reductive amination of alpha-ketoglutaric acid with N-13 labeled ammonia in a reaction catalyzed by the enzyme glutamate dehydrogenase, which was immobilized on an activated Sepharose support (29). After removing unreacted N-13 ammonia by ion exchange chromatography and filtering for sterilization, the labeled L-glutamate was injected intravenously.

Imaging was begun three to five minutes post-injection, since over 90 percent of the administered radioactivity has left the blood by that time. The instrument used in these studies was a dual-detector rectilinear scanner equipped with tungsten and lead shielding and specially designed collimation for purposes of quantitative imaging (18, 31). Data were recorded on computer compatible magnetic tape and analyzed using a system of computer programs which permitted a variety of image processing options, including pixel-by-pixel image arithmetic and absolute quantification of organ or tumor radionuclide content (29, 32). The images produced with this system are two-dimensional, depth-independent and longitudinal in nature. Patients were scanned prior to the initiation of chemotherapy, which generally consisted of a combination of HDMTX, adriamycin, bleomycin, cyclophosphamide and actinomycin-D (6). Imaging with N-13 glutamate was repeated at intervals during the treatment regimen. At the end of chemotherapy, the primary tumor was amputated, or resected *en bloc*, and subjected to a thorough histologic evaluation. The response of the tumor was assigned a grade of I through IV based on its histological response (6); grades I and II indicated incomplete response to the treatment, with many viable tumor cells persisting, while grades III and IV indicated a good response to treatment, with only scattered foci of viable tumor cells, or no viable tumor cells, remaining in the specimen. These results were correlated with the quantitative change in N-13 content in the total tumor volume, relative to the pre-treatment study.

Based on a careful point-by-point assessment of tumor response in a select group of patients (23), a threshold value of 40% decrease in glutamate uptake was chosen to distinguish non-responding tumors from responding ones; that is, if the decrease in glutamate during chemotherapy did not exceed 40% of the pre-treatment value, or actually increased, then the response based on the scan

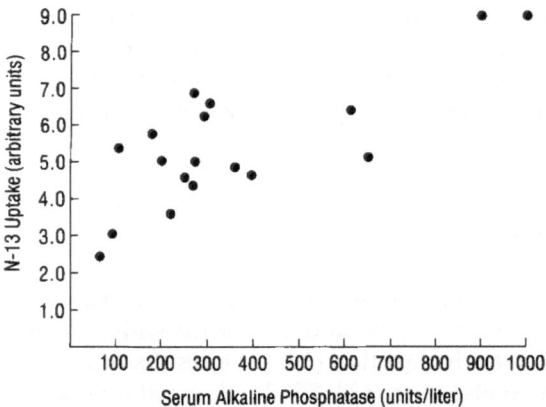

Fig. 1. Relationship between N-13 uptake in untreated primary osteogenic sarcoma following the administration of N-13 L-glutamate and blood serum alkaline phosphatase levels. N-13 uptake was determined via computer analysis of digital rectilinear scanner images. The positive correlation reflects the association between two processes occurring in the primary tumor: the extraction of labeled glutamate from the blood by the soft-tissue component of the lesion, and the accelerated rates of bone mineral turnover in the area of the tumor

was judged to be incomplete. The uptake of N-13 by the primary tumor was also correlated with the blood serum alkaline phosphatase levels at the time of the scan in osteogenic sarcoma patients whose blood levels of that enzyme were elevated.

The relationship between N-13 uptake by the primary tumor and the serum alkaline phosphatase levels in 18 untreated patients with primary tumors in the distal femur is shown in Figure 1. The tumor N-13 uptake is expressed in arbitrary units which are proportional to the fraction of administered radioactivity. The correlation is quite good, considering that the normal values for alkaline phosphatase vary greatly according to the individual's sex and age, which ranged from 6 years to 29 years in this population. In general, the change in N-13 uptake during subsequent treatment reflected the change in alkaline phosphatase levels.

The relationship between the reduction in N-13 uptake and the histological response in 27 patients is shown in Table 1. The results of this study, where the N-13 uptake change reflects changes integrated over the total tumor region, are similar to the findings when small areas of the image and surgical specimen were considered in six patients (23). In both studies, the uptake was found to decrease dramatically where tumor tissue was responding to chemotherapy, based upon clinical observation and eventually upon the degree of necrosis seen microscopically.

Although the change in N-13 uptake by the primary tumor correlates well with the histologic response, the measurement of the uptake using a two-dimensional imaging system suffers from one major draw-back: radioactivity localizing in normal tissues such as skin and skeletal muscle which is interposed between the

Table 1. Relationship between change in N-13 glutamate uptake and histologic response

		Histologic Response Grade[b]	
		Grade I–II	Grade III–IV
N-13 response[a]	Good	2	7
	Incomplete	14	4

[a] A good N-13 response was indicated if a 40% decrease in glutamate uptake was attained during therapy. An incomplete response was indicated by failure to attain 40% decrease, or by increased N-13 uptake

[b] Grades I and II reflect little or no necrosis in examined histologic sections of the resected primary tumor following therapy. Grades III and IV indicate extensive necrosis with little or no residual viable cells detected in the specimen

Chi-square analysis of this contingency table show that the relationship between change in N-13 uptake and histologic response is statistically significant, with $p < 0.01$

tumor and the detectors is included in the image field and must be included in the uptake value, unless some *a priori* knowledge about the extent of the tumor along the axis perpendicular to the imaging plane is available for correction purposes. Since the normal tissue does not change its N-13 concentration during therapy in the same way as tumor tissue, the correlation between response and uptake change may be obscured. The problem of radioactivity concentrating in tissues surrounding the tumor can be overcome by imaging the distribution of N-13 using transaxial PET.

3. Studies with Positron Emission Tomography

The positron emission tomograph used in the studies reviewed here is a commercial prototype based on a design by Brownell and co-workers at the Massachusetts General Hospital (33, 34). It is capable of simultanously acquiring data from 23 contiguous transverse sections in a field of view 30 cm wide. Computer programs were written to rearrange the data in the 23 resulting transverse-plane images into sagittal and coronal sections of any desired thickness. This was accomplished by linearly interpolating the data in the 23 transverse sections along an axis perpendicular to the axis of the planes into 64 sections. The resulting $64 \times 64 \times 64$ image matrix was then addressed to yield the desired coronal or sagittal plane. This procedure, which facilitates the assessment of anatomical relationships in the transverse views, is illustrated in Fig. 2. Figure 3 shows comparative N-13 glutamate and F-18 sodium fluoride images in a twenty-year-old woman with a mesenchymal chondrosarcoma in the left proximal tibia. Fluorine-18 is incorporated into bone by displacing hydroxyl or bicarbonate ions in the crystal lattice at the surfaces of accessible hydroxyapatite crystals (35), and thus reflects bone mineral turnover and formation of new bone as well as tumor vascularity. The agents show similar uptake patterns, except that the fluorine distribution seems to include areas of reactive bone not seen in the glutamate

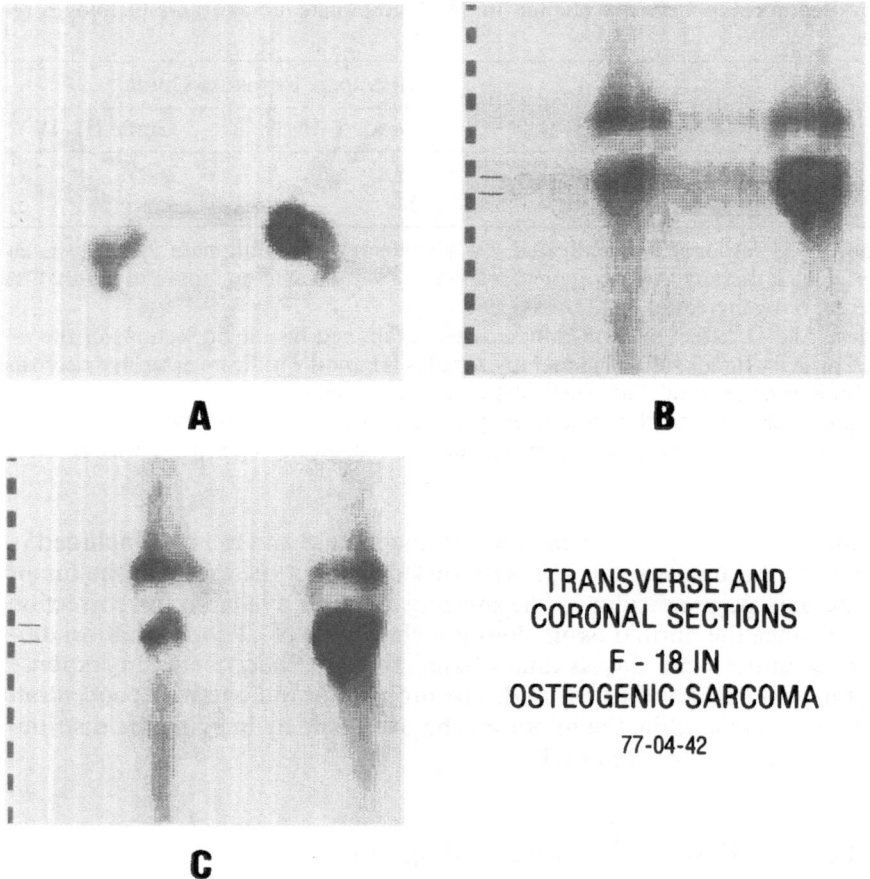

Fig. 2. Procedure for rearranging transverse section data into coronal or sagittal projections. **A** Single transverse section from PC-4200 scanner, showing F-18 distribution in proximal tibiae of 11-year-old female with osteogenic sarcoma. Lesion with increased F-18 uptake is seen in the left tibia. "Streak" artifacts due to limited angular sampling are apparent. **B** The 23 transverse sections are linearly interpolated to the scale of the transverse matrix (64 × 64) and are reprojected onto the coronal plane. Bands due to projection of reconstruction artifacts reduce the contrast in the image. **C** Coronal section image obtained after removal of artifacts by suppressing transverse section counts that are less than an operator-selectable fraction of the maximum counts in a section. Improved contrast is noted. Resolution in the sagittal and coronal planes along the axis perpendicular to the plane of the transverse section images ("z-axis") is improved by correcting each transverse section for the contribution from activity localized in adjacent planes. This is accomplished by blurring the adjacent planes with a filter reflecting the z-axis point-response function, and subtracting the resulting image from the plane of interest in a point-by-point fashion

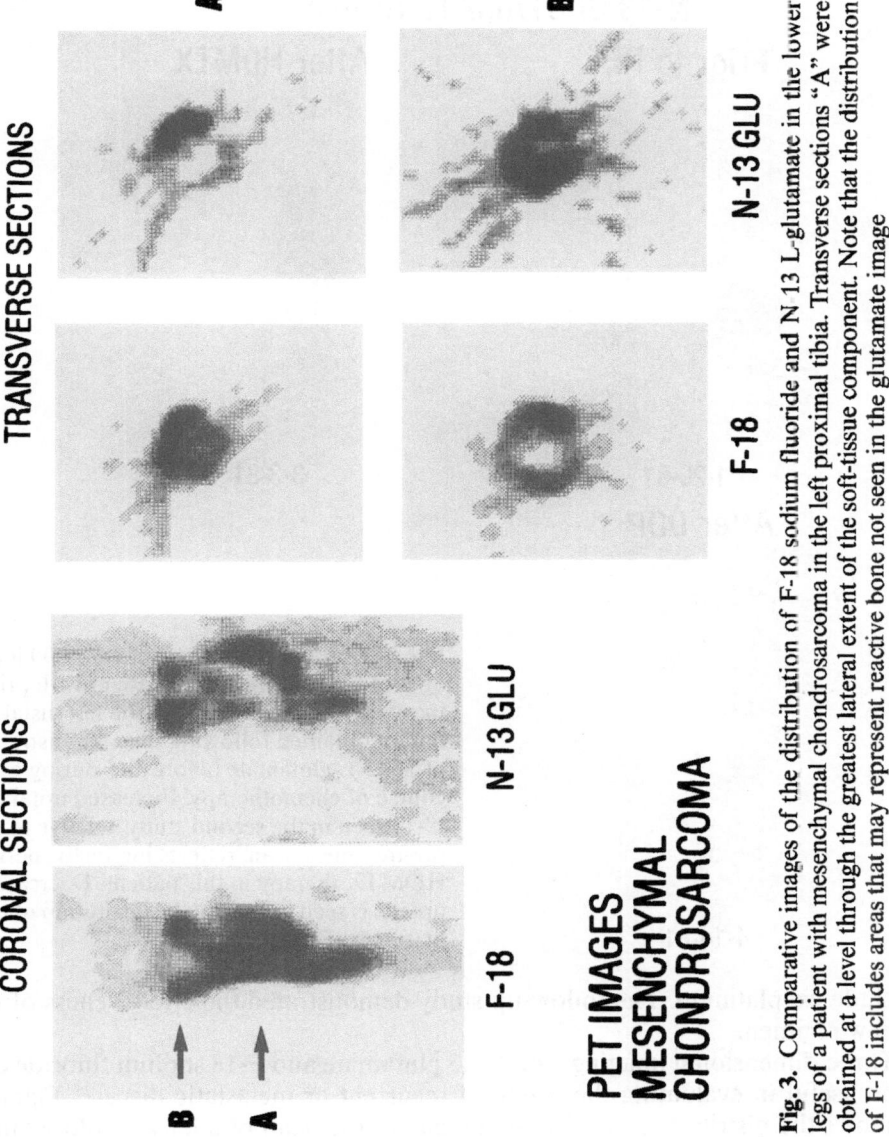

Fig. 3. Comparative images of the distribution of F-18 sodium fluoride and N 13 L-glutamate in the lower legs of a patient with mesenchymal chondrosarcoma in the left proximal tibia. Transverse sections "A" were obtained at a level through the greatest lateral extent of the soft-tissue component. Note that the distribution of F-18 includes areas that may represent reactive bone not seen in the glutamate image

distribution. The anterolateral extension of the soft-tissue mass is best appreciated in the glutamate image. Figure 4 shows serial PET scans in a nine-year-old girl with osteogenic sarcoma who was studied repeatedly with N-13 glutamate during chemotherapy. Following HDMTX therapy, the lesion was seen to concentrate more N-13 glutamate relative to the pretreatment baseline study, and it was concluded that the HDMTX therapy was ineffective. Based on this finding and confirmatory clinical findings, the treatment was changed to in-

N-13 GLUTAMATE SCANS

Prior to Rx   After HDMTX

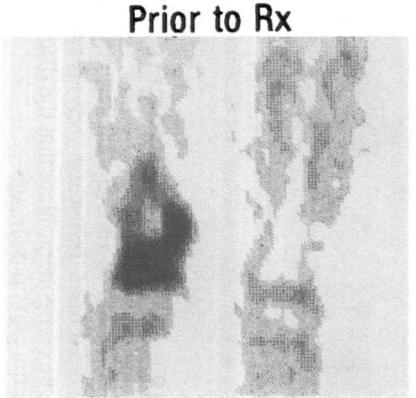   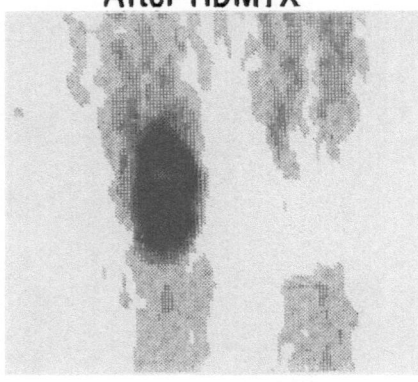

1-20-81   3-3-81

After DDP

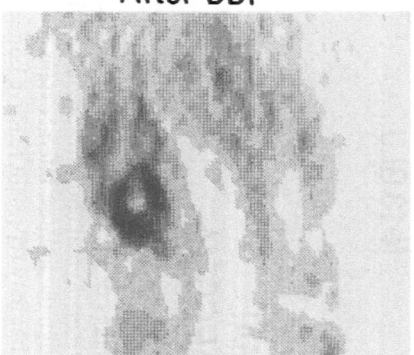

Fig. 4. Coronal projections of PC-4200 images of the lower legs of a patient with primary osteogenic sarcoma in the left distal femur, obtained following the administration of N-13 L-glutamate before and during a course of chemotherapy. Increased uptake of N-13 seen in the second study, relative to the pre-treatment scan, reflects ineffectiveness of HDMTX therapy in this patient. Decreased uptake is seen following the institution of cis-platinum (DDP)

4-14-81

clude cis-platinum. The follow-up study demonstrated the effectiveness of the new regimen.

Three-dimensional imaging with N-13 glutamate and F-18 sodium fluoride can be useful in evaluating the extent of recurrent or metastatic disease. Figure 5 shows the distribution of N-13 glutamate in the head of a 43-year-old woman with recurrent mesenchymal chondrosarcoma arising from the occipital bone. The transverse section, which was recorded at the level of the lateral ventricles, shows extensive invasion of the tumor into the brain. Here, the tumor-to-normal tissue contrast is quite high, due to the fact that N-13 glutamate, a negatively-charged dicarboxylic acid at physiological pH, does not cross the intact blood-brain barrier (26). Figure 6 shows a thin sagittal section through the tumor, which was reconstructed from data contained in the 23 transverse sections. Figure 7 shows a coronal section, reconstructed at the level of the spine, in a 30-year-old woman with metastatic stromal cell sarcoma of uterine origin who was imaged with F-18. Abnormally increased uptake is evident in the posterior

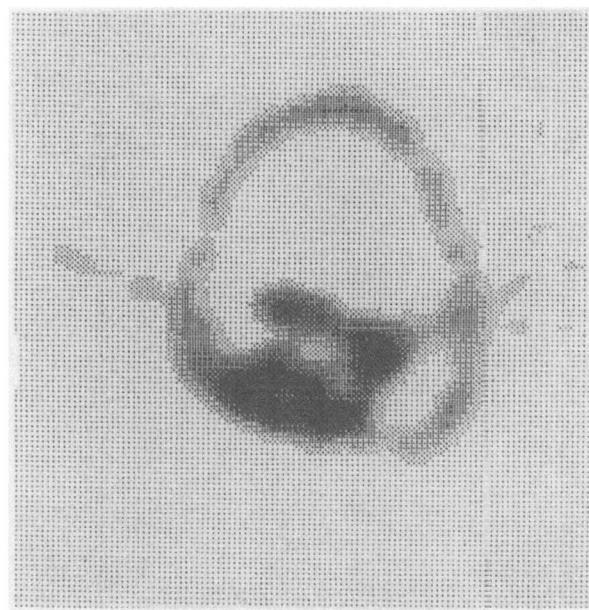

Fig. 5. Transverse section through head of patient with recurrent mesenchymal sarcoma of the occipital bone, obtained with N-13 L-glutamate. Extensive invasion of the brain by tumor is apparent

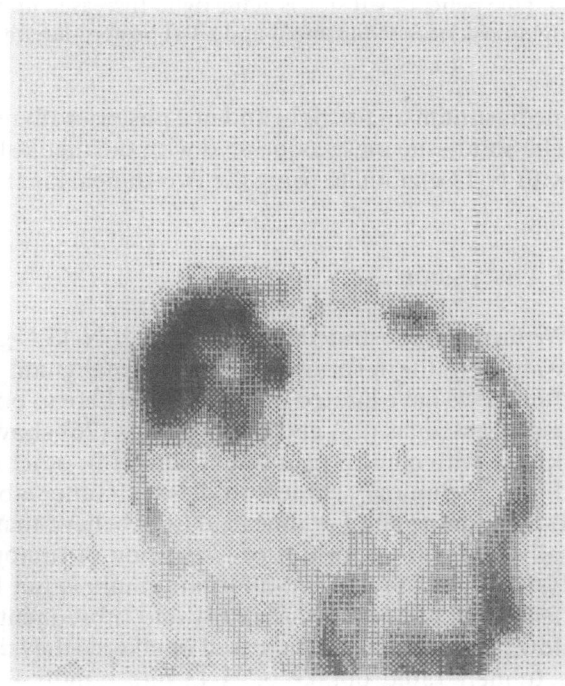

Fig. 6. Thin sagittal section reprojected from the transverse section data for the patient shown in Figure 5. Reprojection of the data allows an increased appreciation of the anatomical relationships among structures seen in the transverse section images

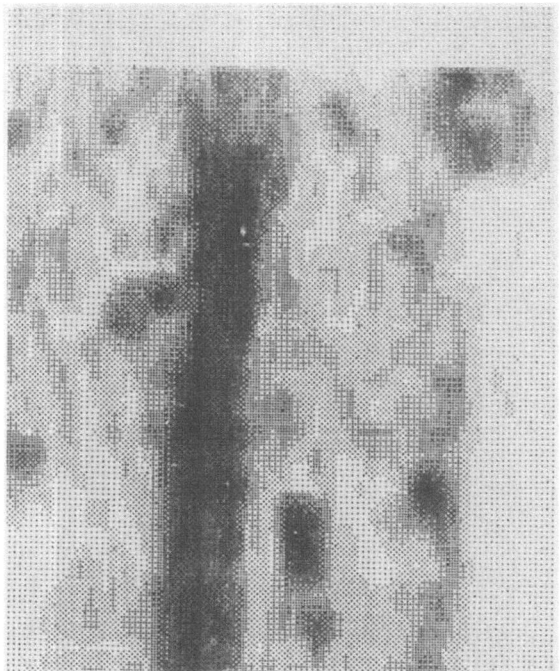

Fig. 7. Coronal projection of F-18 scan of the thoracic region of a patient with metastatic stromal cell sarcoma. In this view, the cephalad direction is at the top of the image, and the patient's left side is at the right. Increased uptake is seen in metastatic foci in the right posterior mediastinum and a left postero-lateral rib

mediastinum on the right and in a posterior rib near the body margin on the left. The soft tissue mass associated with the mediastinal lesion was well-visualized with N-13 glutamate using the two-dimensional HEG system.

4. Discussion

We have demonstrated the potential of N-13 L-glutamate and PET scanning in evaluating the response of primary malignant disease to chemotherapy. Although we have limited our discussion here to bone tumors, our work with soft-tissue sarcomas (25) and brain tumors (26) shows that the correlation of N-13 uptake with clinical response is similar in other types of tumors as well. In addition, preliminary clinical studies with other labeled N-13 amino acids such as L-methionine and L-valine lead us to believe that these agents would be universally useful for purposes of evaluation of tumor response. The mechanism of N-13 glutamate concentration in tumors is not known, although we have evidence that tumor neovascularity and subsequent increase in blood volume and flow play a role in the increased uptake, as they do in F-18 localization (36). The extremely rapid uptake of N-13 into tumors following the administration of

N-13 L-glutamate (90% of maximum typically attained within 45 seconds) makes it unlikely that the tumor is extracting some N-13 labeled metabolite rather than N-13 glutamate itself. Following its extraction from the blood, the N-13 of glutamate can become involved in a number of biochemical reactions in the tumor tissue, including transamination to other amino acids and incorporation into proteins. Studies to determine the metabolic fate of glutamate and other N-13 labeled substrates in tumor tissue are in progress (37).

In conclusion, N-13 labeled amino acids, as well as other physiologic substrates labeled with C-11, O-15, K-38 or F-18, and PET imaging have the following implications for the oncologist involved in clinical research, and for the imaging specialist responsible for developing and implementing diagnostic techniques for the evaluation of therapy:

1. The method provides a means of evaluating the response of the primary tumor to chemotherapy in a quantitative, three-dimensional fashion. The results of the study can provide a basis for continuing an effective course of therapy or discontinuing a regimen that is not effective.

2. The method involves the use of an agent that is physiologically significant, and does not rely on parameters such as tissue density differences that may be secondary to the presence of viable malignant cells. Extraction of N-13 labeled amino acids reflects natural metabolic or neovascular phenomena characteristic of neoplastic tissue.

3. The use of N-13 labeled amino acids potentially provides the research oncologist with powerful tools for studying the *in vivo* or *in vitro* nitrogen metabolism of tumors. Alterations in amino acid metabolism that are known to exist in certain tumors, such as tyrosine and phenylalanine in melanoma and neuroblastoma, could be studied with the appropriate labeled amino acids. The information gained could be exploited to develop new strategies for treatment of malignant disease.

The use of this method requires a PET scanner and a cyclotron, and as a means of evaluating therapy would directly benefit only those patients being treated at major medical facilities with access to such technology. However, research carried out in the course of studying these patients may provide oncologists in other settings with improved drugs, treatment regimens and the techniques necessary to evaluate their effectiveness.

References

1. Jaffe N (1972) Recent advances in the chemotherapy of metastatic osteogenic sarcoma. Cancer 30; 1627–1631
2. Jaffe N, Frei E, Traggis D, Cassady JR, Watts H, Fuller RM (1977) High dose methotrexate with citrovorum factor in osteogenic sarcoma – Progress Report II. Cancer Treat Reports 61; 676–679
3. Rosen G, Tan C, Sanmaneechai A, Beattie EJ, Marcove R, Murphy ML (1975) The rationale for multiple drug chemotherapy in the treatment of osteogenic sarcoma. Cancer 35; 622–630

4. Sutow WW, Sullivan MP, Fernbach DJ (1975) Adjuvant chemotherapy in primary treatment of osteogenic sarcoma. Cancer 36; 1598–1602

5. Rosen G, Murphy ML, Huvos AG, Guttierez M, Marcove RC (1976) Chemotherapy, en bloc resection and prosthetic bone replacement in the treatment of osteogenic sarcoma. Cancer 37; 1–11

6. Rosen G, Marcove RC, Caparros B, Nirenberg A, Kosloff C, Huvos AG (1979) Primary osteogenic sarcoma: the rationale for preoperative chemotherapy and delayed surgery. Cancer 43; 2163–2177

7. Rosen G, Caparros B, Mosende C, McCormick B, Huvos AG, Marcove RC (1978) Curability of Ewing's sarcoma and considerations for future therapeutic trials. Cancer 41; 888–899

8. Rosen G (1978) Primary Ewing's sarcoma: the multidisciplinary lesion. Int J Radiat Ocol Biol Phys 4; 527–532

9. Rosen G, Nirenberg A, Juergens H, Caparros B, Huvos AG (1980) Response of primary osteogenic sarcoma to single-agent therapy with high-dose methotrexate with citrovorum factor rescue. In: Nelson JD, Grassi C (eds) Current chemotherapy and infectious disease. Proceedings of the Eleventh International Congress of Chemotherapy and the Nineteenth Interscience Conference on Antimicrobial Agents and Chemotherapy. Washington DC. American Society for Microbiology

10. Marcove RC, Mike V, Hajek JV, Levin AG, Hutter RVP (1970) Osteogenic sarcoma under the age of twenty-one. A review of one hundred and forty-five operative cases. J Bone Joint Surg [Am] 52; 411–423

11. Smith J, Heelan RT, Huvos AG, Caparros B, Rosen G, Urmacher C, Caravelli JF (1982) Radiographic changes in primary osteogenic sarcoma following intensive chemotherapy. Radiological-pathological correlation in 63 patients. Radiol 143; 355–360

12. Frankel RS, Jones AE, Cohen JA, Johnson KW, Johnston GS, Pomeroy TC (1974) Clinical correlations of Ga-67 and skeletal whole-body radionuclide studies with radiography in Ewing's sarcoma. Radiol 110; 597–603

13. McKillop JH, Etcubanas E, Goris ML (1981) The indications for and limitations of bone scintigraphy in osteogenic sarcoma: a review of 55 patients. Cancer 48; 1133–1138

14. Goldstein H, McNeil BJ, Zufall E, Treves S (1980) Is there still a place for bone scanning in Ewing's sarcoma? J Nucl Med 21; 10–12

15. Cohen MB, Spolter L, McDonald NS, Cassen B (1972) Enzymatic synthesis of N-13 L-glutamine. J Nucl Med 13; 422

16. Lembares N, Dinwoodie R, Gloria I, Harper P, Lathrop K (1972) A rapid enzymatic synthesis of 10-minute N-13-glutamate and its pancreatic localization. J Nucl Med 13; 786

17. Gelbard AS, Clarke LP, McDonald JM, Monahan WG, Tilbury RS, Kuo TYT, Laughlin JS (1975) Enzymatic synthesis and organ distribution studies with N-13-labeled L-glutamine and L-glutamic acid. Radiol 116; 127–132

18. Laughlin JS, Weber DA, Kenny PJ, Corey KR, Greenberg E (1964) Total body scanning. Br J Radiol 37; 287–296

19. Gelbard AS, Christie TR, Clarke LP, Laughlin JS (1977) Imaging of spontaneous canine tumors with ammonia and L-glutamine labeled with N-13. J Nucl Med 18; 718–723

20. McDonald JM, Gelbard AS, Clarke LP, Christie TR, Laughlin JS (1976) Imaging of tumors involving bone with N-13-glutamic acid. Radiol 120; 623–626

21. Rosen G, Gelbard AS, Benua RS, Laughlin JS, Reiman RE, McDonald JM (1979) N-13 glutamate scanning to detect the early response of bone tumors to chemotherapy. Proc Am Assoc Cancer Res, March 1979; 189

22. Gelbard AS, Benua RS, Laughlin JS, Rosen G, Reiman RE, McDonald JM (1979) Quantitative scanning of osteogenic sarcoma with nitrogen-13-labeled L-glutamate. J Nucl Med 20; 782–784
23. Reiman RE, Huvos AG, Benua RS, Rosen G, Gelbard AS, Laughlin JS (1981) Quotient imaging of N-13 L-glutamate in osteogenic sarcoma: correlation with tumor viability. Cancer 48; 1976–1981
24. Reiman RE, Rosen G, Gelbard AS, Benua RS, Laughlin JS (1982) Imaging of primary Ewing sarcoma with N-13 L-glutamate. Radiol 142; 495–500
25. Sordillo PP, Reiman RE, Gelbard AS, Benua RS, Magill GB, Laughlin JS (1982) Scanning with N-13 glutamate: assessment of the response to chemotherapy of a patient with embryonal rhabdomyosarcoma. Am J Clin Oncol (CTT) 5; 285–289
26. Reiman RE, Benua RS, Gelbard AS, Allen JC, Vomero JJ, Laughlin JS (1982) Imaging of brain tumors following administration of L-[N-13] glutamate. J Nucl Med 23; 682–687
27. Yeh SDJ, Benua RS, Grando R, Graham MC (1980) Fluorine-18 positron emission tomography of bone lesions. J Nucl Med 21 (6); P56
28. Yeh SDJ, Myers WG, Grando R, Reiman RE and Benua RS (1982) Carbon-11 dioxide imaging in bone tumors. In: Raynaud C (ed) Nuclear Medicine and Biology. Pergamon Press, pp 1992–1995
29. Gelbard AS, Benua RS, Reiman RE, McDonald JM, Vomero JJ, Laughlin JS (1980) Imaging of the human heart after administration of L-(N-13)-glutamate. J Nucl Med 21; 988–991
30. Moses J, Borer JS, Gelbard A, Reiman R, Devereaux R, Graham M, Lamonte C (1982) Myocardial glutamate metabolism in man: relation of uptake to left ventricular function in aortic regurgitation. Clin Res 30; 208A
31. Clarke LP, Laughlin JS, Mayer K (1972) Quantitative organ uptake measurement. Radiol 102; 375–382
32. Clarke LP, Maugham EZ, Laughlin JS, Knapper WH, Mayer K (1976) Calibration methods for measuring splenic sequestration by external scanning. Med Phys 3; 324–327
33. Brownell GL, Durham CA (1974) Recent developments in positron scintigraphy. In: Hine GJ, Sorenson JA (eds) Instrumentation in nuclear medicine. New York. Academic Press, pp 135–159
34. Carrol LR (1978) Design and performance characteristics of a production model positron imaging system. IEEE Trans Nucl Sci NS-25 (1); 606–614
35. French RJ, McReady VR (1967) The use of F-18 for bone scanning. Br J Radiol 40: 655–661
36. Laughlin JS, Benua RS, Gelbard AS, Reiman RE et al (1981) Report on compounds labeled with N-13 or C-11 used in cancer metabolic studies with quantitative two-dimensional scanning and PET tomography. In: Medical radionuclide imaging (vol II). Vienna. IAEA, p 249
37. Rosenspire KC, Gelbard AS, Cooper AJL, Schmid FA, Roberts J, Young CW (1982) Uptake and metabolic fate of N-13 ammonia and glutamine in glutaminase sensitive and resistant murine tumors. J Nucl Med 23 (5); p 37–38

Labeling of Organic Compounds with Positron Emitters

F. Helus and W. Maier-Borst

Introduction

This review will focus on the interdisciplinary overlap of the physical sciences in the applications of antimatter in radiopharmaceutical research and development.

The primary mission of radiochemistry in combination with bioscientific research can be defined as follows: "to discover, develop and produce radioactive materials as radionuclides or labeled compounds, for direct use in biochemical research and clinical diagnostics, to obtain information from inside the human body without surgical intervention". Renewed interest in the production, synthesis and use of radiotracers and labeled compounds came about as a result of greater availability of accelerators and cyclotrons to biomedical investigators, who recognized the utility of labeling metabolites with short-lived positron emitters of their natural elements. During the same period, the rapid expansion of imaging techniques led in a special way to the development of modern positron-emission transaxial tomography. Using the annihilation coincidence technique, the localisation and reconstruction of the volume element in space where the point of annihilation occured, is possible from outside the body. Using this principle, more can be added to our knowledge of human physiology. Recent developments show that the radiochemical work – preparation of suitable labeled metabolites for the study of physiological processes – will be the most important factor in this field for the next decade. Positron-emitting radiopharmaceuticals can be used in combination with tomographic techniques for the in-vivo study of local metabolism. This is the reason why metabolic tracers labeled with short-lived radionuclides (^{11}C, ^{13}N, ^{15}O, etc.) have gained great importance.

Most biological matter (about 96%) consists of carbon, nitrogen, oxygen and hydrogen. The chart of nuclides, however, shows that radionuclides of these elements have only a limited number of isotopes with externally detectable radiation and useful half-lives: carbon-11 (20.3 min), nitrogen-13 (9.97 min), oxygen-15 (122 sec) and oxygen-14 (70 sec). These neutron deficient nuclides – positron emitters with short half-lives – provide the means to synthesize a wide variety of labeled compounds which are of biomedical interest. Besides simple labeled organic molecules like ^{11}CO, ^{11}CO$_2$, ^{13}N-N, ^{13}NH$_3$, H$_2$^{15}O, etc., more complex organic compounds are used extensively. A complete listing of carbon-11 compounds can be found in reviews of Wolf (1, 2). A review that can serve as a primer on nitrogen-13 and oxygen-15 labeled compounds, has been

Table 1. Important short-lived positron emitters

nuclide	half-life	reaction 1	target material	reaction 2	target material
^{11}C	20 min	$^{14}N(p, \alpha)^{11}C$	N_2	$^{11}B(d, 2n)^{11}C$	B_2O_3
^{13}N	10 min	$^{16}O(p, \alpha)^{13}N$	O_2	$^{12}C(d, n)^{13}N$	$C; CH_4$
^{15}O	2 min	$^{14}N(d, n)^{15}O$	N_2	$^{15}N(p, n)^{15}O$	enr. ^{15}N
^{18}F	110 min	$^{20}Ne(d, \alpha)^{18}F$	Ne	$^{18}O(p, n)^{18}F$	enr. ^{18}O
^{34m}Cl	32 min	$^{34}S(p, n)^{34m}Cl$	NiS		

written by Straatman (3). In addition to the "biological" elements named above, the positron-emitting short-lived halogen nuclides have to be considered (4, 5). Cyclotron produced fluorine-18 (109.8 min), chlorine-34m (32.2 min), bromine-75 (97 min) and iodine-121 (2.12 h) are from a nuclear point of view preferable to the most commonly available I-131 and I-125. Table 1 shows the important short-lived positron emitters with their half-lives and appropriate target material being used.

In this field, in-vivo characterisation of tumor pathophysiology at the cellular and molecular levels has become of major diagnostic interest. Studies of the mechanisms by which tumors compete with normal host tissues for essential nutrients are especially important for the investigation of tumor growth (6). Sophisticated experiments designed for such investigations must begin with studies of local substrate utilisation. The quantification of regional perfusion for the assesment of nutrient's availability to the region of interest, which exhibits several stringent requirements for the tracer to be employed, is a part of this problem.

Experimental Design and Methods

Instrumentally it is a very long way for the radioactive tracer from the place of its origin to its delivery as a radiopharmaceutical ready for application. This complex preparation procedure has, if a short-lived radionuclide is involved, to be terminated during an extremely short period of time. The radioisotopes of the "truly biological" elements, whose radiation is measurable outside the patient body, all have half-lives of less than half an hour; so transport from production source to patient is a limiting factor. To overcome this time barrier, cyclotrons have been installed in several medical centers around the world.

Radiochemistry as science includes basic research, determination of excitation functions, isotope production and processing, the rapid preparation of labeled intermediates, target technology, hot atom chemistry and the study of chemical effects of nuclear transformations. Figure 1 shows the position of radiochemistry between the pure physics which represents the cyclotron as a producer of accelerated particles, and medicine as the field of application of radiotracer in radiopharmaceutical form.

The choice of a new radionuclide or new labeled compound for research in medicine or biochemistry depends on a combination of physical, chemical and

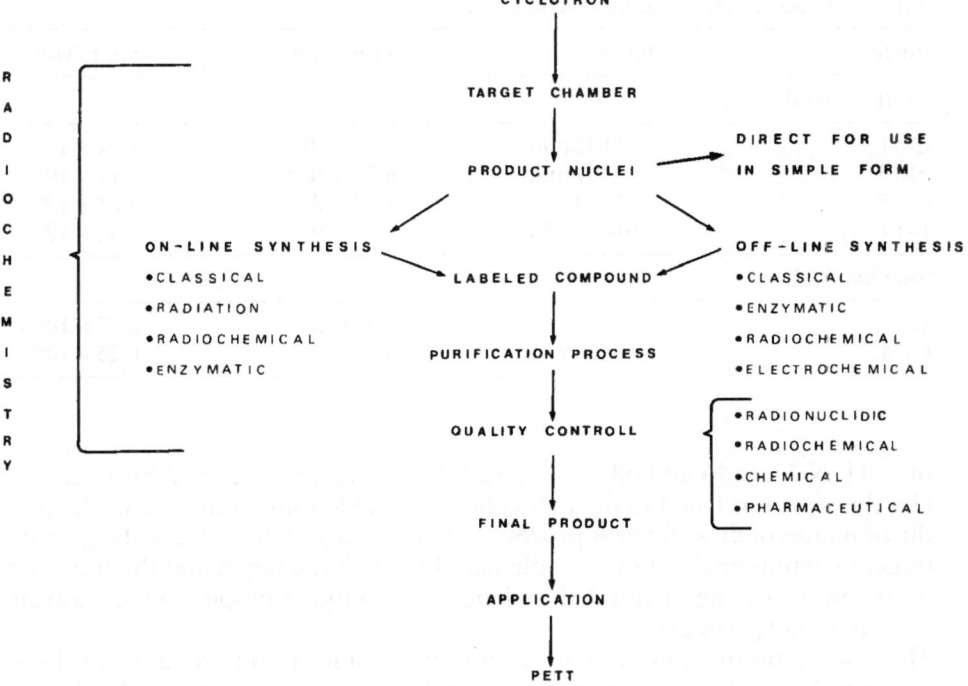

Fig. 1. The position of radiochemistry in bioscientifical research work

physiological considerations. The interdisciplinary cooperation is an absolute requirement for the production of short-lived biological tracers. The fast procedures for labeling complicated biological molecules need a deep knowledge in organic chemistry, especially in the modern fast labeling techniques (use of catalysts and enzymatic synthesis). In planning a synthesis several factors have to be considered:

1. place of the radioactive label in the molecule,

2. specific activity,

3. amount of radioactive product desired.

The synthesis itself can be either inorganic, organic, biochemical or electrochemical (see Fig. 1.) Some aspects of the method used are similar to those used for the preparation of C-14 compounds. However, the preparation of carrier-free or no-carrier-added (NCA) compounds, including the problems associated with carrier contamination, are completely different. Other differences are related to the half-life of the radionuclides and the limited number of radioactive precursors with which the preparation can be started.

We have studied the production techniques and irradiation parameters, and have developed and constructed special target systems for routine production

Table 2. Specific activity of important nuclides

nuclide	half-life	1 mCi (mg)	Ci/mMol
short lived labels			
C-11	20.32 min	1.19×10^{-9}	9.24×10^{6}
N-13	9.97 min	6.89×10^{-10}	3.77×10^{7}
O-15	122.10 sec	1.62×10^{-10}	1.85×10^{8}
F-18	109.72 min	1.05×10^{-8}	3.42×10^{6}
long lived labels			
H-3	12.33 y	1.04×10^{-4}	5.77×10^{1}
C-14	5730.00 y	0.224	6.25×10^{-2}

of C-11, N-13, O-15 and other radionuclides in a simple chemical form. Generally the reactions listed in the first column of Table 1 are preferred, as the production rates of these nuclear processes are generally higher. Use of the gaseous target materials enable more simple handling with the target and the transport of the products. High natural abundance of the target isotope is a great advantage for target gases used.

The routine production of short-lived radionuclides (C-11, N-13, O-15, F-18, etc.) for chemical use and research in our laboratory started a couple of years ago and some biologically useful compounds have already been synthesized and used in preliminary experiments. A very important factor for the production of cyclotron radionuclides is the target technique. From the point of view of handling the activity, gas targets are preferable as compared to solid and liquid targets. Fortunately all "biological" positron emitters may be produced from gaseous target material and recovered as a gaseous product nuclide. Convenient target chambers have been constructed for routine production of C-11, N-13, O-15 and F-18 in our laboratory. Regarding the target development, small target volume and stopped-flow irradiation technique have been found to give higher specific activity products. This implies that the cross-section of the beam should be small.

The nuclear reaction yield of product nuclei depends mainly on the beam current of accelerated particles. The nuclear reactions induced with charged particles produce neutron deficient radionuclides, mostly resulting in product nuclides different from target nuclides. This implies that theoretically "carrier-free" high specific activity products can be prepared. Table 2 shows the theoretical specific activity of four short-lived positron radionuclides in comparison to the specific activity of long-lived H-3 and C-14 which are used in biochemical studies and in-vitro analysis.

Carrier-free labeled compounds may be of relevance in biomedical research because high specific activity may be added to biological systems without significantly changing the physiology or biochemistry of the stable element or compound already present. Another advantage may be, that even very toxic compounds can be applied. However, in practice true carrier-free conditions are difficult to achieve, because the risk of contamination with stable nuclides during the production of the radioactive nuclei and the preparation of the labeled compound is hard to avoid. Carbon is so abundant in nature that practically every contact with organic material brings inactive carbon as carrier into the product, which generally results in lower specific activity of carbon-11 labeled compounds. Dilution of the labeled compound with inactive carbon as additional carrier may decrease as the number of synthetic steps and additional chemicals used decrease. Strict precautions concerning pollution of the product by atmospheric $^{12}CO_2$ in the target gas and all reagents have allowed a maximum specific activity for various prepared compounds of at best 1 to 2 Ci/μmol, at the time of use. Fluorine-18 labeled compounds can be obtained with much higher specific activity if there are appropriate chemical procedures, which, with fluorine, are not easily available every time. Using nuclides with a very short half-life is sometimes rather difficult. In complicated synthetic procedures the time factor has to be considered carefully for all synthetic steps. A rule of thumb is that the compound must be available for application within three half-lifes after the termination of radionuclide production. On the other hand, repetitive applications may be possible within a short period of time.

Carbon-11 as a label for organic mulecules continues to occupy an important place in biomedical studies and is therefore chosen as a representative example for description of our activity in this field. The number of options for the application of C-11 (20.3 min) seems to be unlimited by the variety of components of biological material. Its use covers physiology, biochemistry, toxicology and pharmacology. Its half-life period is sufficient to allow complex molecules to be synthesized using ^{11}CO, $^{11}CO_2$, $H^{11}CN$ and $^{11}CH_3I$ as precursors. Thus a number of radiopharmaceuticals such as amino acids, polyamines, sugars, hormones, fatty acids and drugs can be labeled. Different approaches to the labeling of complicated organic molecules were reviewed by Wolf and Redwanly (2).

Carbon dioxide labeled with C-11 is produced via the ^{14}N (p, a) ^{11}C reaction by irradiation of pure nitrogen gas with 14.0 MeV protons using the German Cancer Research Center Compact Cyclotron. To avoid all organic joints, the target is made entirely from aluminium, equipped inside with a quartz liner. The welded inlet aluminium foil (1.1 mm thickness) reduces the incident energy of protons from 21.0 MeV to 14.0 MeV. The total beam energy absorbed by nitrogen gas under 11 kp/cm² pressure amounts to 12.5 MeV. With a beam current of 15 to 20 μA and an irradiation time of 30 to 40 minutes, about 700 to 900 mCi of $^{11}CO_2$ are produced. Production parameters are given in Table 3.

After irradiation, $^{11}CO_2$ is trapped on a small aluminium trap pre-cooled with liquid argon at a flow rate of about 500 cm³ min⁻¹. The $^{11}CO_2$ is dispensed by heating the trap with a hot air fan and passing a stream of carrier helium through it into another reaction vessel. More than 90% of the activity is transferable by the procedure described. Radiogaschromatography is used to analyze the chemical composition of the irradiated gas. Typically, more than 95% of the final product is found to be in $^{11}CO_2$ form; the remainder is almost entirely ^{11}CO. The co-production of ^{13}N, over the ^{14}N (p, pn) ^{13}N reaction, may be suppressed by diminuation of the initial proton energy in the target to 14.0 MeV.

Table 3. Production parameters for the $^{14}N/p$, $\alpha/^{11}C$ reaction. Trapping efficiency and specific activity of the carbon-11 dioxide produced

Target gas	N_2	O_2	CO/CO_2	Hydrocarbons/CH_4/
composition:	99.9995%	0.00005%	0.00001%	0.00001%
	other impurities: H_2O, H_2; Ar; 0.00051% in total.			
inactive CO_2:	3.93×10^{-7} mg cm^{-3}			
pressure:	11 kp cm^{-2}			
throughput:	600 cm^3 min^{-1}; total amount of irradiated gas during irradiation: 18 000 cm^3			
total inactive CO_2 from gas impurities:	7.1×10^{-3} mg			
absorbed energy:	12.5 MeV, lost from incident 14.8 MeV			
total charge:	3.6×10^4 µAsec; average beam current 20 µA			
yield $^{11}CO_2$:	2.59×10^{10}–3.3×10^{10} Bq (700–900 mCi) in total, trapped in 1N NaOH solution. 1.6×10^{10}–2.59×10^{10} Bq (450–700 mCi) trapped on molecular sieve 4 Å			
trapping efficiency:	64–78%			
recovery of $^{11}CO_2$:	77–92%			
inactive CO_2 transferred:	5.1×10^{-3} mg (at maximum efficiency and recovery conditions)			
specific activity of $^{11}CO_2$ for synthesis:	5.56 Ci/µmol			

^{11}C labeled carbon dioxide as a parent precursor can easily be transformed into other important precursors, all of which belong to a family of chemically related compounds. A precursor is defined as the simple starting compound for chemical synthesis. It may be obtained directly from the target, or derived, mostly on-line, from products formed in the target.

As $H^{11}CN$ and $^{11}CH_3I$ were identified as important intermediates in the synthesis of amino acids, sugars and other compounds, we have studied the fast synthetic methods for its preparation and have developed a remotely controlled procedure for the on-line production of $H^{11}CN$ from $^{11}CO_2$, the primary nuclear reaction product. Some published data are available describing methods for the synthesis of $H^{11}CN$ (7, 8, 9, 10). Our study is based on the industrial process of HCN production, known as the BMA-process (11). For the synthesis of $H^{11}CN$, the nickel catalyzed reduction of $^{11}CO_2$ to $^{11}CH_4$ is readily achieved with hydrogen at 420 °C. $^{11}CH_4$ is formed selectively and no other hydrocarbons are detected quantitatively in the reactor effluent. In the second catalytic reaction $^{11}CH_4$ is oxidized on the platinum catalyst at 1000 °C with NH_3. The setup for the routine production of $H^{11}CN$ is shown in Fig. 2.

The optimal production parameters for the synthesis of $^{11}CH_4$ and $H^{11}CN$ were determined in terms of both chemical and radiochemical yields. Table 4 summarizes the experimental parameters and the yield of each reaction step. The data were determined by radiochromatographic analysis.

The development and construction of the apparatus shown in Fig. 2 was a precondition for the direct and on-line preparation of the basic stones for the synthesis of complicated organic molecules with high yields.

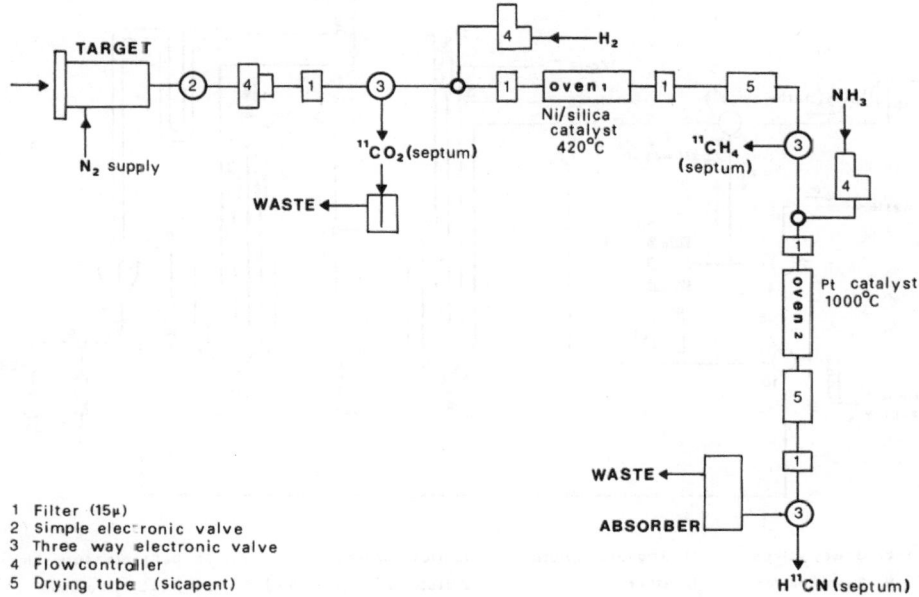

1 Filter (15μ)
2 Simple electronic valve
3 Three way electronic valve
4 Flow controller
5 Drying tube (sicapent)

Fig. 2. Apparatus for the on-line preparation of $H^{11}CN$

Table 4. Yields and experimental parameters of the catalytic conversion of $^{11}CO_2$ to $H^{11}CN$

reaction	yields (%)				
	$^{11}CH_4$	$H^{11}CN$	^{11}CO	$^{11}CO_2$	Σ
$^{14}N(p, \alpha)^{11}C$			5.08	81.23	86.31[a]
$CO_2 + 4H_2 \rightarrow CH_4 + 2H_2O$ $\}$ 95.42				0.8	96.22[b]
$CO + 3H_2 \rightarrow CH_4 + H_2O$					
$CH_4 + NH_3 \rightarrow HCN + 3H_2$	4.1	60.16	29.87	1.9	96.03

furnace parameters:

product	heated tube length (cm)	inner diameter (cm)	T (°C)	catalyst used
$^{11}CH_4$	5	1.5	420	500 mg Ni, supported on Kieselgur (55–60% Ni, 100 m²/g between quartz wool plugs)
$H^{11}CN$	20	0.4	1000	1 g Pt wool, pressed to an actual length of 0.5 cm

rates of flow:

target gas	H_2	NH_3	
600	10	10	ccm/min

[a] the difference to 100% implies the coproduced ^{13}N activity
[b] the difference due to formation of hydrocarbons

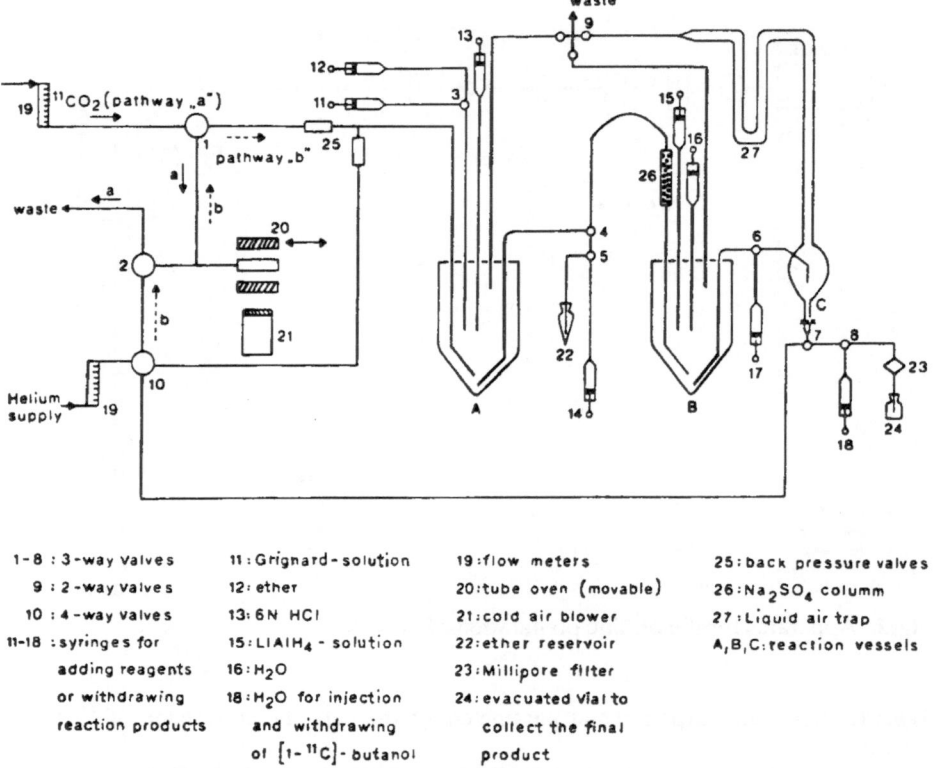

Fig. 3. Apparatus for the preparation of [1-^{11}C]-butanol

1-8 : 3-way valves	11: Grignard-solution	19: flow meters	25: back pressure valves
9 : 2-way valves	12: ether	20: tube oven (movable)	26: Na$_2$SO$_4$ columm
10: 4-way valves	13: 6N HCl	21: cold air blower	27: Liquid air trap
11-18 : syringes for	15: LiAlH$_4$ - solution	22: ether reservoir	A,B,C: reaction vessels
adding reagents	16: H$_2$O	23: Millipore filter	
or withdrawing	18: H$_2$O for injection	24: evacuated vial to	
reaction products	and withdrawing	collect the final	
	of [1-^{11}C]- butanol	product	

Table 5. Synthesis of [1-^{11}C]-butanol

(a) $CH_3CH_2CH_2MgCl + {}^{11}CO_2 \rightarrow CH_3CH_2CH_2{}^{11}COOH$

(b) $CH_3CH_2CH_2{}^{11}COOH + LiAlH_4 \rightarrow CH_3CH_2CH_2{}^{11}CH_2OH$

In the course of our development of metabolic tracers labeled with short-lived radionuclides, we have synthesized carbon-11 labeled butanol, for application as a tracer in the assessment of regional blood supply. The generally known synthesis of [1-^{11}C]-butanol has been adopted and semi-automated preparation has been evaluated for no-carrier-labeling with carbon-11. Table 5 describes the two steps reaction mechanism for the synthesis of [1-^{11}C]-butanol.

The synthesis is based on the carboxylation of n-propylmagnesium chloride with $^{11}CO_2$ and subsequent reduction of the resulting free [1-^{11}C] butyric acid with LiAlH$_4$ dissolved in anhydrous ether. The detailed synthesis and the results have been published previously (12). Carbon-11 labeled butanol is now routinely available as a tracer for biomedical experiments. The scheme of the set-up used for the routine production of [1-^{11}C]-butanol is shown in Fig. 3.

As a part of our development of metabolic tracers labeled with short-lived radionuclides, the synthesis of following substances is presently being developed:

1. 1-^{11}C-2-Deoxy-D-Glucose and 2-Fluoro-2-Deoxy-D-Glucose

2. l-aminocyclobutane carboxylic acid (ACBC) and

3. l-aminocyclopentane carboxylic acid

4. α-methylornithine

D-Glucose is the major energy source of normal brain. Interest in glucose metabolism has prompted several investigations to prepare ^{11}C-Glucose and ^{11}C-Glucose analogs. The chemical preparation of 1-^{11}C-2-Deoxy-D-Glucose has been accomplished by the Brookhaven group using Na^{11}CN as precursor (13, 14). A review by Vaalburg covers the general literature of carbon-11 labeled compounds through 1982 (15). It is sometimes very difficult or impossible to reproduce some of the published works. Many articles do not contain relevant information, the omission of which makes them less useful than they could be. Biomedical research requires a reliable routine production of labeled tracers. Therefore, proper research and preliminary studies of certain syntheses are absolutely necessary. Presently we are studying the preparation of ^{11}C labeled 2-deoxy-D-glucose (16) and ^{18}F labeled glucose analogues (17). For the fluorination of sugars and other organic molecules, special anhydrous fluorine-18 in no-carrier-added conditions has to be prepared; this is another of our research problems which have just been studied.

Carboxyl-labeled ^{11}C-DL-amino acids are prepared from ^{11}CO$_2$ and α-lithiated isocyanides and by amination of α-bromo acids and by the Strecker synthesis (18, 19). The unnatural alicyclic α-amino acids l-aminocyclobutane carboxylic acid (ACPC) and l-aminocyclopentane carboxylic acid (ACBC) (20, 21) were obtained by the same procedure. ACPC and ACBC have been studied and synthesized according to the modified Strecker synthesis as described by Hayes and Washburn (20). The synthesis occurs via a spirohydantoin intermediate, starting with the ketones cyclopentanone and cyclobutanone, and with ^{11}C-labeled sodium cyanide in presence of ammonium carbonate. The hydantoin formed initially, is hydrolysed using a strong alkaline solution in a pressure vessel at a temperature just below the decomposition temperature of the resulting acid. The proper time for synthesis and hydrolysis of the hydantoin has to be adapted to the short half-life of C-11. Thus, a compromise between overall synthesis time and maximum yield related to the starting activity of carbon-11 labeled cyanide (^{11}CN)$^-$ is established. The purification of the free amino acids ACPC and ACBC is carried out on pretreated ion exchangers. The total synthesis time is about 35 minutes for the preparation and 40 minutes for the separation. Thus, for a reasonable activity of 0.3 to 0.5 mCi/cm^3 required for animal experiments, an initial activity of about 1.0 to 1.5 Ci of (^{11}CN)$^-$ is required in order to overcome the long synthesis time and bad chemical yield of about 20.0 to 25.0%.

Summary

During the past decade, chemists have synthesized a variety of compounds in which short-lived positron emitters such as C-11, N-13, O-15 and F-18 were incorporated (1, 2, 3, 15). These compounds are designed to enable the research on various physiological parameters. These include blood flow, blood volume, oxygen and glucose metabolic rates, drug-receptor interactions, protein synthesis, amino acid utilisation, permeability of blood-brain barrier and others. Positron emission tomography studies are expected to lead to a better understanding of disorders such as cancer, epilepsy, heart diseases, stroke, senile dementia and mental illnesses like schizophrenia. By the great number of publications presented by the leading scientific institutions, it is documented that there is intensive research work in this field. It is very well known that when new promising methods are established, they will also generate many difficulties, not least of which is the problem of cooperation between different scientists. Further development of radiochemistry, in general, has its fundament in the biological or medical use of product tracers and their application in research in physiology and biochemistry. The new trend and new tasks must be based on the application of new ideas, new and fast synthetic procedures such as enzymatic, catalytic and biomimetic synthesis (22, 23).

A discussion of the potential uses and production of radionuclides and labeled compounds in the broad spectrum of biomedical research is obviously beyond the scope of one short paper. Instead, we have presented a couple of examples of experimental work which has been done in our laboratory, and described production procedures for positron emitting radionuclides which have been delivered on a fairly regular basis.

Experimental work in this field requires an enormous amount of intellectual and material resource. It is necessary for scientists to go together and have lively contacts and cooperation on European and world bases, in order to run the research properly and to get the best cost – benefit relationship.

References

1. Wolf AP et al (1973) Synthesis of Radiopharmaceuticals and labeled compounds using short-lived isotopes. In Radiopharmaceuticals and labeled compounds. Vol 1. Vienne. IAEA, pp 345–381
2. Wolf AP, Redvanly CS (1977) Carbon-11 and radiopharmaceuticals. Int J Appl Radiat Isotopes 28; 29–48
3. Straatman MG (1977) A look at ^{13}N and ^{15}O radiopharmaceuticals. Int J Appl Radiat Isotopes 28; 13–21
4. Palmer AJ, Clark JC, Goulding RW (1977) The preparation of Fluorine-18 labeled radiopharmaceuticals. Int J Appl Radiat Isotopes 28; 53–65
5. Stöcklin G (1977) Bromine-77 and I-123 radiopharmaceuticals. Int J Appl Radiat Isotopes 28; 131–147
6. Johnstone RM, Scholefield PG (1965) Amino acid transport in tumor cells. Adv Cancer Research 9; 143–226
7. Finn RD et al (1971) The preparation of cyanide-^{11}C for use in the synthesis of organic radiopharmaceuticals II. Int J Appl Radiat Isotopes 22; 735–744

8. Lamb JF, James RW, Winchell HS (1971) Recoil synthesis of high specific activity ^{11}C-cyanide. Int J Appl Radiat Isotopes 22; 475–479

9. Christman DR et al (1973) Production of carrier-free H^{11}CN for medical use and radiopharmaceutical syntheses. J Nucl Med 14: 864–866

10. Christman DR et al (1975) The production of ultra high activity ^{11}C-labeled hydrogen cyanide, carbon dioxide, carbon monoxide and methan via the ^{14}N (p, a) ^{11}C reaction. Int J Appl Radiat Isotopes 26; 435–442

11. Endter F (1958) Die technische Synthese von Cyanwasserstoff aus Methan und Ammoniak ohne Zusatz von Sauerstoff. Chemie-Ing Tech 30; 305–310

12. Oberdorfer F et al (1982) The synthesis of 1-^{11}C-Butanol. Radiochem Radioanal Letters 53; 237–252

13. Shiue CY et al (1978) The synthesis of 1-^{11}C-D-Deoxy-2-Glucose for measuring regional brain glucose metabolism in vivo. J Nucl Med 19; 676–677

14. Fowler JS et al (1979) Agents for the armamentarium of the regional metabolic measurement in vivo via metabolic trapping: ^{11}C-2-Deoxy-D-Glucose and halogenated derivatives. J Lab Radiopharm 16; 7–9

15. Vaalburg W, Paans AMJ (1983) Short lived positron emitting radioisotopes. In: Helus F (ed) Radionuclide Production, Vol II. CRC Press, Florida, pp 47–103

16. Ehrin E et al (1983) C-11 labeled glucose and its utilization in positron-emission tomography. J Nuc Med 24: 326–331

17. Ido T, Wan CN, Fowler JS et al (1977) Fluorination with molecular fluorine, a convenient synthesis of 2-deoxy-2-fluoro-D-glucose. J Org Chem 42; 2341–2342

18. Washburn LC et al (1976) C-11 labeled aminoacids for pancreas visualisation. J Nucl Med 17; 557–558

19. Weinges K et al (1971) Die externe asymetrische Strecker-Synthese von α-Methylaminosäuren. Chem Ber 104; 3594–3606

20. Hayes RL et al (1978) Synthesis and purification of C-11 labeled aminoacids. Int J Appl Radiat Isotopes 29; 186–187

21. Washburn LC et al (1979) 1-aminocyclobutane-^{11}C-carboxylic acid, a potential tumour seeking agent. J Nucl Med 20; 1055–1061

22. O'Donnell MJ, Eckrich TM (1978) The synthesis of amino acid derivatives by catalytic phase-transfer alkylations. Tetrahedron Letters 47; 4625–4628

23. Breslov R (1979) Biomimetic chemistry in oriented systems. Isr J Chem 18; 187–191

Subject Index

Positron Emission Tomography of the Brain

Editors: **W.-D. Heiss, M. F. Phelps**

1983. 108 figures. XVIII, 224 pages
ISBN 3-540-12130-7

The contributions by noted international researchers contained in this book provide a critical survey of the scientific and clinical capability of positron emission tomography for measuring cerebral metabolism. The authors review the principles involved, the methodology for regional quantification of isotope content in brain tissue, and the isotopes and labeled compounds used. Techniques and clinical results of regional blood flow measurement, oxygen uptake and glucose metabolism are extensively reported, and new developments and initial results of quantification of regional protein synthesis and receptor distribution are indicated.

Early Detection of Breast Cancer

Editors: **S. Brünner, B. Langfeldt, P. E. Andersen**

1984. 94 figures. XI, 214 pages
(Recent Results in Cancer Research, Volume 90)
ISBN 3-540-12348-2

The contributions presented in this book deal with all aspects of the diagnosis, classification, pathological anatomy, and treatment of breast cancer. Of special interest are reports on the results of screening projects in Sweden and the USA, on progress in early diagnosis using ultrasound and subtraction techniques, and on the current use of irradiation and breast-preserving surgical intervention in place of mastectomy in the USA and Denmark. The book is a synthesis of valuable information from scientists at leading mammographic centers, and will be useful to diagnosticians, radiotherapists, and surgeons as well as to medical students.

Springer-Verlag
Berlin
Heidelberg
New York
Tokyo

Manual of Clinical Oncology

Edited under the auspices of the
International Union Against Cancer
3rd fully revised edition. 1982. 44 figures.
XV, 346 pages
ISBN 3-540-11746-6

The continuing success of the UICC's *Clinical Oncology,* the refinement of their educational objectives for medical students and young practitioners, and the significant advances scored over the last few years in cancer research have all led to the decision to publish a third edition of this manual. It presents students and practitioners with a concise summary of essential oncological knowledge. Although emphasis is placed on basic aspects, the clinical imlications are made clear.

"... this well-prepared compact volume fulfills admirably its role as an aid to the understanding of malignant disease for all sections of the medical community."

British Journal of Surgery

UICC
International Union Against Cancer
Union Internationale Contre le Cancer

TNM-Atlas

Illustrated Guide to the Classification of
Malignant Tumors
Editors: **B. Spiessl, O. Scheibe, G. Wagner**
1982. 311 figures. XII, 229 pages
ISBN 3-540-11429-7

The *TNM-Atlas* is an illustrated guide to the UICC's system for classifying malignant tumors, a system that has been adopted worldwide since introduction of the definitive version in 1978. The atlas is designed to facilitate documentation of clinical findings by combining the precisely graded TNM classifications with topographical drawings of the organ or anatomic section under consideration. All physicians actively engaged in clinical oncology will find that this book provides the ideal framework within which to conduct daily anamnesis.

Springer-Verlag
Berlin
Heidelberg
New York
Tokyo